THE
COSMOS
SUPREMACY

PRAMODINI SHETTY

INDIA • SINGAPORE • MALAYSIA

Notion Press

Old No. 38, New No. 6
McNichols Road, Chetpet
Chennai - 600 031

First Published by Notion Press 2019
Copyright © Pramodini Shetty 2019
All Rights Reserved.

ISBN 978-1-68466-491-7

DISCLAIMER

This is a work of fiction, and resemblance to any place, thing, situation or character, dead or alive, is purely coincidental. Certain basic facts have been incorporated to lend a touch of realism to an otherwise imaginary narrative.

***

DEDICATION

This book is dedicated to my daughter Akanqsha Shetty for being the reason for my living, and my parents, my father, the late Sunder Hegde and my mother Shakuntala Hegde who are the reason for my existence.

* * *

ACKNOWLEDGEMENTS

I would like to express my heartfelt thanks to my daughter Akanqsha Shetty for being the greatest support in my life. Without her this book would never have been published.

***

CHAPTER I

Grrrrrrrrrrrr... His feet froze. For a split second fear crippled his cerebrum catapulting it to insensibility. The moment was ephemeral, labile, charged and indecisive, but not his instinctual defence mechanism which was, at all times, alert. Run? Or stand still holding his breath, in a bid to fool the vicious brute into thinking there was no human presence around? It was crucial to weigh his options astutely, if he had to see the dawn of another day. The endless existential canards that had been dormant for a long time transuded through his mind, and memories of his father's bereavement came flooding back. His harsh unexpected demise had shaken the entire locality to shocking disbelief. It was ironical that a man, as brave as he, had been so easy a prey to a fiendish four legged creature.

The incident had cast an ominous spell on the village. There were no similar encounters thereafter on this 'once upon a time' trodden path. And yet, it continued to instil fear in the minds of the inhabitants. The passage of time had done nothing to dent the panic. Most of them avoided the foredoomed track totally. The bolder among them, however, continued to use it, but they were far too few in number and were considered fatuous by those who believed that they would soon meet the fateful end that his father had.

For a brief while he too had sidestepped it, but pinning down his sense of adventure hadn't been easy. Soon he resumed using it. As far as he could recall, no one had ever heard the intimidating growl of the cruel animal let alone have an interlude with it, not since the death of his father 3 years ago. This was a first. But now he was unsure. Had he been too casual and unwise in dismissing the on and off reports of damp paw prints? Had the lure of this narrow, crooked side track which was an artist's delight been a devil's temptation? Was there a possibility, remote as it seemed to be, that the sharp canined bully was still pillaging around?

His pulse quickened, his chest convulsed and his heart beat faster.

The roar died down. Shaking himself vigorously to get rid of the oppressiveness that enveloped the surroundings, he brushed off a dry leaf that had fallen on his shoulders and quickly walked on. "Just my imagination," he muttered as a wave of relief swept through him.

He relaxed.

His age was controversial. Sixteen? Or Eighteen? Most people presumed that the latter number sat aptly on his shoulders. Not his mother. She insisted he was 16. The misperception prevailed because there were no written records at the time of his birth. His was an age that had been calculated on the basis of the seasons that passed by.

Whatever the argument, there was no denying that he was young - way too young, to supervise employees far older than he. Demanding total obedience from seniors who didn't always take kindly to the thought of reporting to a person several years younger to them, was indeed onerous. But his unpretentious grit under the toughest of situations helped him sail through. Yes! The job was arduous and had its fair share of jerks and bumps, but then what was life without a few extra kicks and starts, or so he believed.

Adolescents, in this part of the world, were not necessarily encouraged to work, and yet, contradictorily, there was an unspoken belief that age should never be a constraint for those who wanted to earn a living. The industrious and the energetic were appreciated, while those who shirked responsibility were frowned upon.

The Master had, initially, been reluctant to employ this youngster, but having seen sparks of his brave father in him, he was confident that he would deliver. And deliver he did, with such exemplary ease and alacrity that he had no reason, whatsoever, to regret his decision, or even reconsider it. The lad neither belied his expectations, nor betrayed the trust reposed in him. The cherry on the cake was his perceptiveness–a quality not easy to find in this tiny village 'Likwi.'

'Likwi'–the name was queer. Nobody knew why it was called so, but that was what it had been known as for as long as anyone could remember. Quaint and unspoilt, this charming hamlet consisted of a cluster of houses with red tiled roofs, a few make shift kiosks that sold home grown produce and a teensy grocery shop that stocked basic provisions like cereals, pulses, vegetables and toiletries. The only excitement that relieved the boredom of its humdrum existence was the weekly village fair.

The village or town, as some preferred to call it, had seen no consequential growth for a long, long time. Even in these modern times, it continued to be untouched by the gross commercialisation that had infected the rest of the world. Its only touch to the outside world was a more sophisticated department store situate almost 2 miles from its epicentre. It displayed and sold luxurious items like clothes, shoes, perfumes, chocolates and electronic items.

Gattu lived in a modest dwelling perched at the edge of a tiny rivulet which overflowed during the monsoon but ran dry in summer. The humble sequestered abode had a roof made of some gravelly substance, and lay separated from 'Likwi' by a thick jungle. Deep woods on 3 sides added to its mystery and allure. Nature was strikingly beautiful here with a wisp of scented air wafting all the year round, irrespective of the weather.

Two routes led to the village from this place. The longer one was frequently walked upon. It was preferred because it was considered safe even though it took a person almost 60 minutes to cut across to the village, from here. The stubbier one was just a thirty minute trail, and yet it was painstakingly avoided as it traversed through a dense, impenetrable, dark forest. Presence of wild animals in the thickest part of the section was not the only reason why it was shunned.

Hush whispers of ghastly creatures gliding through the lush woods surreptitiously in search of human flesh was another excuse to avoid it. This was the route on which his father had been mauled, and it was this fortuity that fuelled the trepidation of the occupants further.

The short track was picturesque and breathtakingly beautiful. Faint rays of light filtering in sparse fragments through the heavy foliage of the tall trees shrouded it in suspense. Architected by nature, the variables that cropped up on this path were invigorating. The scent of fresh green leaves, the earthy smell of moist mud, the barely perceptible macabre sounds that sent a frightening but thrilling chill through him, the thick vegetation, the strange berries that he feasted upon, and the singing of the crickets.....were all exciting excerpts that added to the joy of living... his living. There was a curious comfort in passing the same trees, in touching the same branches that dipped in the wind, and in finding his way through the same thickets that transformed themselves astonishingly with the change in season.

Changes in climate brought about a change in the colour of the leaves, the texture of the bark and the softness of the earth, but his comfort... well! It never changed, it just differed in degrees. The ecstasy of living so close to nature and experiencing her dance was incomparable. He empathized with those who deprived themselves of the munificence of Mother Earth.

Many envied the Master's son who attended affluent schools overseas; but not he. Comparisons were odious and he knew that, but he also knew that the vagaries of human nature were not always controllable. It was quirky to liken himself to an individual who lived in a different plane altogether. He never once forgot that Sameer was the son of a well heeled prominent member of society while he was just a humble employee of that household. But there was sinful pleasure in stolen thoughts that made him feel privileged, and he occasionally indulged in such analogy.

Society may have treated him like a pariah, but not Mother Earth. She had been extremely partial to him. Why else would she have singled him out to be her favourite student? Why else would she have hand-picked him to be privy to her well guarded secrets which she chose to share with only the adventurous few? The thought gave him untainted pleasure. He considered himself blessed.

'Master' was a pseudonym for the local landlord who provided employment to most people in and around the village. He had the comforting geniality of influential men who despite being plenteously laden with material assets, stayed grounded and were easily accessible even to the humblest without appearing too condescending. He was extremely popular. His residence, the most distinguished address of the locality, was a resplendent structure standing on a plot with seemingly endless boundaries. It consisted of 3 storeys and 75 rooms. A larger part of this edifice remained unoccupied, except on occasions when it was thrown open to tourists who sought a transitory place to stay in.

The Master's only son, Sameer, was ambitious. He moved in an urbane circle where people dressed their speech with superficial pretensions. The putative cognoscenti sprinkled their savoir faire with conversations on topics like 'the rising population' and its equally growing demand for food, shelter and clothing.

These affected personages whom society dubbed as the 'elite' were painfully shallow. They attempted to outdo each other in voicing their opinions on social issues with great ferocity and equally great fanaticism; not because they cared about transforming society, but because they thought it was fashionable to do so. Populist statements made with a studied carelessness attracted the limelight on them, and gave them a sense of frivolous importance. The ersatz intellectuals, who were arguably low on EQ, boasted of presumably high IQs which were as faux as they. Their concern was superficial, their knowledge contemptible, and their lifestyle, devoid of altruism.

Despite belonging to the self styled cream of society, Sameer was different. He had a strong sense of purposefulness. He was determined to prove to the world that even a nondescript place like Likwi could become self-sufficient; and not just self-sufficient, but perhaps even partially meet the world's demand for food, fuel and fibre, in the near future. Use of hybrid seeds of selected varieties, technologically advanced equipment, and energy subsidies in the form of irrigation water, fertilizers and pesticides, would indisputably increase the farming efficiency of his village.

Yes! 'Likwi' could easily become a 'food hub' of the world. It had the land, it had the people and it had the natural resources. All it needed was an exposure to 'modern' agriculture and the right application of it. It was a lofty ambition cocooned with a vehement hope. Many who fawned in his presence ridiculed him behind his back, but he remained unruffled. Derision and criticism were to be taken with a pinch of salt, if one had to succeed in one's goals.

His own place of birth lacked the global exposure or the expertise to assist him in accomplishing his vision—a vision of turning this modest little village or town, into a modern 'agricultural' destination. And to achieving this end, he enrolled himself in a University abroad.

The education standards in 'Likwi' were depressing. The village had just one municipal school and the school had just one classroom... a classroom that reeked of emptiness. The students made a mockery of their attendance by showing a distinct preference for the adjoining ground where a crude form of football was always being played. The dwellers displayed little or no interest in mastering the 3 R's. The truly literate could be counted on one's finger tips.

It was no mean task trying to impress upon the autochthons the need to be well versed in reading and writing. The younger ones felt asphyxiated in the close confines of the 4 walls and the older ones thought it a sheer waste of time, but the school teacher was neither dissuaded nor discouraged by the indifference of the denizens. Instead, their attitude only served to strengthen his resolve. Transforming the face of literacy of this place was his goal, and he pursued it with dogged determination. He trusted his ability and remained untiring in his standpoint. All that was needed was a right attitude with a right approach. His conviction was an offshoot of his own creed that a strong will always succeeded, when all else failed. He strove to encourage literateness by using imaginative ways.

His several attempts included playing a game of football with the children, and utilising moments in the field to subtly explain the merits of the written word. His efforts were admirable, but not valued. The Master, perhaps, was the only one who motivated him. He fortified the teacher's efforts by accompanying him in his exertions, whenever he had some time at his disposal. "Don't you worry," he would say, "someday you will meet with success."

Gattu was an exception. He loved school, but the gruelling sessions at the Master's house restricted him from pursuing his craving for learning and satiating his thirst for knowledge. At the end of each working day all that his tired body desired was a good night's rest, to rejuvenate and energise itself for the following day's duties. He did, however, try to develop a few latent skills and improve upon his linguistic abilities by visiting the schoolmaster whenever he had a few odd moments to spare.

Reminiscing, he strode down the chemin leading towards his home. Ordinarily he would have walked at a more relaxed pace sinking in the scents and sounds of nature, but not today. He was late. There had been last minute chores to attend to; little things that had been overlooked while handling more important tasks, and not to forget the false alarm in the form of a tiger's growl, temporarily paralysing him on his tracks. His weary body looked forward to the hot meal that would be waiting for him.

The Sun had gone down and the sky above looked abnormally grey. In the distance he heard the dulcet sound of the nightingale, the melodious cooing of the cuckoo and the relaxing sound of the hermit thrush. One never tired of the strains of these wonderful winged critters. They transported Heaven to Earth. Humming as he walked along, he tried to recollect the song the school teacher had taught him recently. What was it? Well, it went something like this...

Kookaburra sits in the old gum tree,

Merry, merry king of the bush is he.

Laugh, Kookaburra, laugh Kookaburra,

Gay your life must be!

Or was it? He was unsure. The words? Were they correct? Had he blundered, yet again? Was his pronunciation impeccable? Or did he fumble once again with his accent? Had he struck the right notes? Or was he off key? He was uncertain.

Of late, he had been making a tremendous effort to familiarise himself with the English language - a language he found interesting, but rather amusing. Why was the word 'put' which was pronounced as 'put' written as 'p u t' while the word 'foot' which rhymed with 'put' written as 'f o o t' instead of 'f u t.' Honestly it was so warping; even thinking of it fatigued him mentally. He scratched his head in confusion. It was no wonder then, that Indian traditions had always stressed upon having a 'Guru' who would guide students along the right path. 'Likwi' was lucky to have one in the form of the present school teacher whose sincerity towards his profession was laudable. He made a mental note to meet him the next morning to seek his reassurance.

A fresh rumbling of thunder broke his reverie. His eyes gave an upward glance and surveyed the expanse above. Grey sullen clouds! Densely laden with water - a sign of an impending heavy rainfall. A streak of lightning that flashed across the sky was followed by a resounding sound. The slight drizzle through which he now walked had superseded the intense shower that had earlier broken down on to the Earth. He was wrong in surmising that the drizzle would peter into a comparatively drier weather; apparently it was just a momentary respite before another rainstorm reared its ugly head. He had to move faster if he wanted to avoid being caught in it.

Quickening his pace, he forced his feet to take longer strides. He was tired and to keep his torpor at bay, he switched from humming to whistling as he skimmed off the tiny droplets of water that kept falling on him.

Night had set in, and it was dark. The bright yellow green flashes of the fireflies dispelled the darkness intermittently. They were not the only insects around, but they were undeniably the most fascinating; but what was even more captivating was the landscape. It, as always, acquired a refreshingly different dimension as the lush green forest took on a bluish crimsonish tint, and the moon played hide and seek with the silver grey cumulus, in the backdrop.

Suddenly he stiffened. His sixth sense warned him of something amiss. As an afterthought, he realised that the nocturnal sounds had ceased to be - a pointer to the possibility of some peril ahead. These criatures of the Earth who lived by their instincts were the first to sense danger. Nature was man's best teacher and he had, during his annees of growing up, begun to rely on her to guide him.

Born of the earth and having lived so close to it, his intuition had been sharpened to sense the finest of indiscernible details normally overlooked by an ordinary individual. Warily he looked around trying to dismiss the malaise that plagued him. He was unable to assign a reason for his disquiet. He tried to divert his thoughts, but failed. A feeling of discomfort swaddled him.

Was fear lurking in the subconscious or was hunger making him feel so? He strode with accelerated strides as his thoughts meandered on the tricks the dark night played on the human brain. Years of negotiating with illusions of nature had educated him to the fact that fear, as an emotion, controlled humans only if one let it dominate them, not otherwise. But wait a minute! What was that? A mirage? An illusion? He remained transfixed to the ground. Perhaps the curtains of exhaustion were now drawing over his consciousness. This was not real. It had to be a delusion. Rubbing his eyes languidly he looked, and then looked again.

At a distance he noticed a strange blurred reddish tincture which was atypical... also... an ill defined, unanalysable quirky buzzing sound, as if a thousand bees were swarming around. This exotic shade of red was something he had never descried before. He knew that for certain. Life in the wilds had educated him to distinguish the known from the unknown. This colour was eerie, alien and totally bizarre. But what exactly was it? Would it be sensible of him to inch closer to it? Or would it be judicious to ignore it and trudge home?

He disliked hesitancy which made his thoughts vacillate like the pendulum of a clock. It undermined his confidence. He was aware that decisions taken during such conflicting moments didn't necessarily give anticipated results; on the contrary, they were often offset by disappointments. But this weird vista dangerously excited his quest for the unknown. His gut feeling warned him against venturing in that direction, but his snoopiness given to making painstaking estimates of the trivia of his life, egged him on. Throwing caution to the winds, he elected to embark on forbidden territory even though it meant that he would have to move away from his comfort zone.

Encroaching on to an area from where he could vaguely distinguish the red glow, he took a right turn and crossed a thick growth of trees interspersed with tall bushes. He sensed a trickle of blood oozing from the inner part of his left arm where thorns tore against his soft skin as he brushed past the densest part of the jungle. He turned right and then left and then right again... the place was a labyrinth not easy to follow... but he had to move if he had to discover the source. As he moved closer and closer to the reddish hue, it became brighter and brighter.

* * *

Illiteracy had never been an obstacle in her pattern of life. Like most living creatures cradled by nature, she lived by her instincts. The rice porridge continued to simmer in the earthen pot placed over the fire place. The last flames of fire had flickered down to give way to orange glowing embers, exuding warmth to the entire room.

Her anxious eyes looked up at the dark sky sparsely scattered with stars. The shadows cast by the surrounding elements denoted that it was well past his regular time. With the night sounds gradually filling the air, the frown lines on her otherwise smooth forehead became more pronounced. She chose to wait a while longer before venturing forth to seek help of her closest neighbour who lived almost a mile away.

Going there at this hour was not an advisable thing to do, but she was left with no choice. Peering into the darkness, she stretched out her neck to see if she could spot even a semblance of a figure in the horizon. There was none. Suspicion turned to conviction. She felt herself tremble at the presentiment of something grisly to come.

Her head covered by an improvised headgear made of woven dried palm fronds, she trod gingerly on the damp mud, her bare feet caressing the dewy grass. She shuddered as her footsteps stilled against something soft. She reckoned it was a python that may have curled itself into a coil. Leaping over it, she continued to walk.

There were puddles of water everywhere, and the ground was still wet from the earlier precipitation. At a distance she heard the movement of a herd of elephants. And what was that fearful sound? Was that the groan of a hungry tiger or was she just delirious? She was a woman of infinite opinion and under varied conditions, she would have rushed through the questions–Who? Why? What? Where? How? But not today. Her emotions were in a state of flux and she was in no mood to understand the 'whys' and 'whats' of anything. Her only son had not come home, and his whereabouts were of greater importance than anything else.

The neighbour's house, though unpretentious, certainly surpassed her own in terms of amenities. It had quite a few conveniences like electricity and even a pipe line through which water flowed into their house.

"Ahoy! Are you awake?" she yelled.

There was no response. She raised her voice to its highest pitch, "Anyone awake?" she shouted even louder than before. Her soft voice, at its loudest, sounded more like a scream.

The door creaked as it opened slowly from inside. An elderly gentleman, affectionately called Chachu by the villagers, with a shawl thrown carelessly over his shoulders formed a silhouette at the doorway. Wisdom sat squarely on his shoulders.

A man of experience, most villagers sought his advice in times of trouble. He looked at the tiny woman standing outside his house in surprise. Her frail appearance was a deception; it contradicted a valiant character that had weathered greater storms in the past. She was fiercely independent, self-reliant and sought support only when all other efforts failed. He knew she was in dire need of help by the way she fixed her searching eyes on his face. She would never have hazarded forth to seek his aid at this late hour, otherwise.

Displaying genuine angst on his countenance, he asked, "Anything the matter?"

"Yes." She sounded hysterical. "Gattu!'

"Gattu? What about him?"

"He has not reached home, yet." Her quivering lips were the only giveaway of anguished emotions on an otherwise calm face.

"What?" he bawled vociferously as he gave a cursory glance at the sky, "Not reached home? Perhaps he has stayed back at the Master's place. The weather, today, has been really egregious and unpredictable." He made a sweeping gesture with his outstretched hands as if to reaffirm the fact.

"I did think of that," she murmured, "but he has never done it before, and that's what has got me worried."

"True! True!" agreed Chachu. "Gattu has never stayed back at the Master's place ever before, not even under the most unspeakable conditions, but not all days are the same and there are exceptions to every rule. I don't see any reason why you should subject yourself to such restlessness. I'm sure he is perfectly safe and fine."

"I hope so too," she said in a voice that was as unsteady as the feet she stood on.

"Hope? That's too light a word. I'm certain he is," replied Chachu with great self assurance.

Chachu belonged to the category of people who believed that optimism in the face of hardship never led to disappointments. But despite the confidence with which he spoke, he found it strange that his voice sounded hollow. It was hard for him to shake off the feeling of restlessness that had imperceptibly, for no reason at all, started creeping on him, and yet he continued in what he presumed was a stable tone.

"I insist there's no reason for anxiety," Chachu assured her, "and I'm only echoing what I said earlier, when I say that the weather has been exceptionally bad today. In all my past 60 years, I don't recall an instance when there was such clamorous thunder and lightning. I'm sure, the Master may have advised him to stay back as a protective measure. And he can't disobey the Master, can he?"

"No, he can't. Maybe you are right," she admitted wistfully. "I should have thought of that possibility before. How ridiculous of me to agonise so incessantly! I'm sorry! It was very inconsiderate of me to disturb you at this late hour. I better return home now," she said apologetically.

Chachu looked at her with compassion. His visage radiated pity. She was a woman who had always fought her personal battles heroically against all odds.

"Stay here for the night," he advised her as he gave her a soft paternal pat on her back. There was something in his intonation that was immeasurably comforting, and yet it failed to free her of the edginess looming over her. She hoped that her intuitive panic was merely an offshoot of her maternal emotions.

"I don't think I should," she replied.

She found herself mired in conflicting thoughts, but even through the mental fracas one thing was clear; she wanted to be alone within the confines of her unpresuming home at this hour. She had lived there, together with her son, for so long that just his aura, despite his absence, would be adequate consolation for her. Chachu's monotonous words rambled on but she was so engrossed in cogitation that he sounded as if he was addressing her from a distance.

"Be prudent," he was saying. "It would be in your best interests to stay back. The weather is mercurial and the jungle gnarly."

"No," she replied tersely a little shocked at the rudeness in her tone. Her frazzled nerves failed to keep her in control of her sensations. "I don't want to sound defiant," she continued softly, in an attempt to temper her earlier offensive tone, "but I would feel better if I go. He may return home while I am here and worry unnecessarily if he finds me not."

"If he does, he's bound to come here. He is aware that this is the only place you visit other than the Master's."

He was right. Chachu generally was. He had the ability to sense the pulse of the moment and this ability coupled with his genuine involvement with the people around him, always ensured that his predictions were never way off the mark. He continued coaxing her, "I insist you stay; going back alone at this hour would be risky. I'll accompany you in the morning."

She paused as she debated with a moment of cunctation and then replied with a firmness that left no room for argument, "Thank you, but I will have to take your leave now. I feel miserable and staying here will only make me more apprehensive."

She walked back briskly, praying fervently. Frenetic moments like these made her nostalgic. They brought back fond remembrances of her husband. His lack of formal education had never been a deterrent. His demeanour, in the face of fear, had been so contagious that people appeared braver than normal in his presence. While he lived, he did odd jobs for renowned photographers in pursuit of rare wildlife and unusual pictures. He accompanied and guided them through the inner recesses of the deep woods, right up to the minacious undisclosed hideouts of the animals of the wild.

His was the hand that propelled photojournalists to great fame and rich money, while he himself lived in the shadows. At the end of every such expedition he related his spine-chilling escapades to her with such detailing that she actually experienced them as if she had been a part of it. Those moments of togetherness were priceless. They had always made her feel as if she was the pivot around which his life circled.

The monetary compensation that he received was not princely, but it helped them eke an existence where basic necessities were never scarce. Occasionally, though, he was handsomely compensated for his proficiency and skill. Had he lived longer, he would have provided her with a better standard of living, But death snatched him away a little too soon. He had been an ideal husband. The only thing she had grudged him was his love for the wondrous environs. It had kept him out of his house most of the time, so much so, that she had perpetually complained of his injustice to the institution of marriage.

Her heart swelled with pride as her thoughts meandered through the soft tissues of her brain. She recalled how he had been instrumental in healing many unknown diseases by the skilful use of herbs that only he had been acquainted with. His selfless attitude had been 'a pearl of great price' for his innate goodness. His exceptional practical knowledge and the impressive application of it had included extracting poison out of a man's wound when bitten by a venomous reptile.

He had, while he was alive, earned the reputation of 'a miracle man,' - a well deserved title that remained unchallenged to this day. People still spoke about how he had saved the Master's wife from the jaws of death by administering a potent herb found in the most perilous part of the forest. It was stark gratitude that had compelled the Master to offer Gattu a job when his father died while trying to save himself from a man-eater.

Her husband's demise had been a cruel blow that fate had dealt on her. It had encumbered her son with responsibilities at an age when all he should have been doing was enjoying the sheer innocence of life. Of the 3 children that she had borne, he was the only one who had survived to the age of 16. The others had succumbed to death for some reason or other during their infancy. She prayed zealously, clasping her two palms in sheer desperation.

Like his father, Gattu was also very brave. She fervently hoped that no danger had befallen him. His safety was paramount. Chachu's view that he may have stayed back due to bad weather was far-fetched. The explanation was too feeble to offer succour to her harassed mind. She held back her tears resolutely. She had shed enough to last several aeons, the last being on her husband's death. It had been a catastrophic moment when she had shielded her inner incapacitation with a show of bravado.

Deaths were a complete mystery to her. Nature could indeed be cruel if she wanted to. She failed to understand why she was the chosen one for such a sad event. Three deaths within such a brief span of time were unfair, but she had tided over the situation. Copious tears notwithstanding, she had gradually gathered the threads and prepared herself to face the world anew along with her only surviving son Gattu.

The blue-white flash of lightning that streaked across the sky, followed almost immediately by a double clap of thunder, put a brake to her thoughts and urged her to walk faster. The heavy rainfall that would most likely fall sooner than expected, would slash across the unpretentious brick walls and rattle the odd tiles of her roof that was badly in need of repair. Her thin legs cut across the pathway to transport her to her house. She was home before the huge, heavy raindrops touched the ground. She waited for the Sun to rise.

* * *

It was still not daybreak when Chachu set out for Gattu's home. His tough exterior belied his inner sensitivity. He had spent the night consoling himself that sleeplessness was merely a matter of growing old, but he knew that it was not so. His self assuagement had been superseded by his reflections and the subliminal capacity of his mind to deal with skew-whiff situations.

The previous night's encounter with the panic-stricken woman had kept him awake all night long. He had allayed her fears by inducing her into believing that her son had probably stayed behind at the Master's house due to bad weather, but he himself had not been convinced. He knew Gattu well. He had braced harsher climate, in the past, to come home to avoid causing anxiety to his mother whom he dearly loved. Undeniably, the boy was extremely young but he displayed a maturity way beyond his 16 years.

"Has he come?" he asked of the woman who was seated at the doorstep. Her tired eyes had closed involuntarily. She woke up with a jolt on hearing Chachu's voice and gave herself a forceful joggle before standing up a little shakily.

"No." She sounded weak and exhausted.

"Then, I think, we should leave for the Master's house right away; he may be there," he said more for easement rather than affirmation. What he didn't add was that his sangfroid was at its lowest ebb. He, who had always prided himself on his intrepid stance in unforeseeable situations like these, was strangely in need of support himself.

The Master had just dusted off his fingertips, taken a quick swig of tea and picked up the morning newspaper with the intent of perusing it, when he noticed Chachu entering the compound through the front gate. He was not alone. He was accompanied by Gattu's mother. Both he and his wife who were having breakfast, looked up sceptically at them. The Master spoke first, "Anything the matter?" he asked.

"It's Gattu," Chachu replied, "he didn't return home last night. Wondered if he was here."

"No," said the Master, leaving the half-finished cup of tea on the table and standing up. "He left late from here last evening. Something important cropped up at the eleventh hour and it was imperative to attend to it right away. In fact, I was expecting him post lunch so that he could rest well before attending to the more arduous tasks of the day."

The Master took a breather before continuing, "He has been working extremely hard, of late. That boy is an asset. He seems to have a never-ending energy for work and I laud him for it; but human energy needs to be conserved if it has to be put to good use. A tired body compromises with the quality of work, you know."

Chachu shook his head absent-mindedly. He dared not look at the slender, expectant lady standing beside him. Her expression would have unnerved him. Making an ill-concealed attempt to put up a brave front he exclaimed, "Where could he have gone then?"

The Master looked askance at him.

"He didn't return home last night," said Chachu.

The Master got worried. It was not like the boy to act so recklessly. After a moment of hesitancy, he echoed Chachu, "Where could he be then?"

Chachu said nothing.

"Have you searched the jungles?" asked the Master

"No," replied Chachu. "That's the only thing left to do. The possibility of him getting lost there seems quite ridiculous. Nobody knows the terrain as well as he does, but when one has lost out on choices then the wisest thing to do is to opt for the most obvious."

"I agree. Take 2 of my men along; the more, the better. The task will be less strenuous," said the Master.

The men scoured the forest for over 6 hours. The search was a goner. Gattu was not found; neither his body nor even a shred of clothing. He had apparently vanished into thin air.

His mother was inconsolable. Life had always been in a constant state of flux - anxiety, tension and conflict. Situations changed easily from the peaceful to the frenzied. Three years ago, when the death of her spouse had rendered her almost crippled, she had traced and retraced her life holding on to the irreversible moments until reality dawned that she would have to face the world bravely if only for her only child Gattu. And now he too was gone. She shrugged in total resignation failing to veil her helplessness, "What do I do now? My only child has disappeared."

"We'll have to report to the local police station," replied Chachu.

* * *

CHAPTER II

"Under total command?" Maxus asked Pontus as the 'Mono Martian Capsule' surfaced up the Martian territory, swerved on the marked spatial runway and alighted majestically in the space aerodrome.

"Absolutely!" replied Pontus.

"Propitious!" proclaimed Maxus, his otherwise steady voice betraying a trace of excitement. "I feared that it would yet again be a narrative of disillusioned hopes."

"This time it's not," asserted Pontus.

"Thankfully so. How often have we discussed a development of this kind, only to be confronted by a reversal?"

"True," accepted Pontus.

"And to think that it has finally materialised! Solutions, I guess, come with time," recognised Maxus.

"They do," emphasized Pontus. "The "Homework" done would most certainly ensure that. We can't always be thwarted in our mission, can we? But it isn't easy. You can't just enter a Planet, pick up a sample and walk away without causing a flutter. Timing is the key. You get that right; you get everything else right."

Maxus acknowledged the admission, and then added, "But with a qualification. Unforeseen and invisible parameters can most easily put us in a position of being out conceptualised. Technicalities!!! ... They can be dealt with; it's the human angle that's unpredictable. Emotions can never be classified into well defined boxes, can they?"

"Your forebodings are not ill placed," admitted Pontus, "but that is no reason to presume that all ventures are doomed failures. Sometimes, some things abide, and just as well."

"I concur unreservedly! Not *all* ventures are doomed failures," said Maxus with a special emphasis on the word 'all,' "but futile attempts do lead to misgivings. Recall our last challenge, for instance. Did it ever strike us that of all the incapacitations, it would be the Venusians who would put us at a troublesome disadvantage?"

The very mention of the Venusians put Pontus in a fuming rage. Clenching his teeth he sputtered, "Most certainly not! Those bloody meddlesome Venusians!"

"So you can't really impugn me for my wariness, can you?" Maxus clarified.

"Most certainly not," stressed Pontus with a grimace that confirmed the shared scepticism of professional confidences. "But we learn from failures. They expose us to our limitations and weaknesses and bear mentation despite impediments."

"Agreed," acceded Maxus,

Pontus carried on with his recitation, "Let's consider the last one just referred to. Because of it, we now know that it is not enough to merely concentrate on facts, systems and the unpredictability of the human behaviour. It is equally important to factor the existence of other entities in Space. Our stratagem this time embraced that parameter - interference of alien celestial objects."

Pontus paused momentarily, as if seeking a riposte from Maxus.

Maxus gave no reply. Instead, he looked at Pontus inquiringly.

"Now that the hominid tester has been picked up, I sound smug," stated Pontus. "But to be honest, I thought another washout was in store. Despite integrating the confirmed possibility of the fickleness of human comportment, it continues to be the most difficult aspect."

Maxus gave him an acerbic stare.

"We had been monitoring his movements for quite some time now," continued Pontus picking up the threads from where he left off, "and had no doubts, whatsoever, of his time of arrival at the spot, but he failed to turn up. There were moments of anxiety. For a fleeting moment, our earlier abortive endeavour haunted me... but fortune favoured us, and the human side also worked out satisfactorily. He did appear, but much later than anticipated."

Pontus paused, a significant pause. "We succeeded in capturing the apposite situation. It was exact, perfectly executed."

The preliminary exchange being dispensed with, Pontus and Maxus engaged themselves in a recap of the object that had just been lifted from Earth... critical details like age, education and the IQ level. The interchange, at the superficial level, was banal and calm, but the undercurrents were menacing and perilous. This was no light conversation. It had dangerous reverberations. A seemingly simple dialogue between these 2 individuals would have explosive ramifications if all went well, as per plan.

"16 Earth years?" asked Maxus merely to reiterate what was deliberated earlier.

"Yes, approximately 16."

"Approximately?"

"Yes."

"Not sure?"

"No. His birth has not been recorded. He may be a little older, but this is no cause for concern. He is well suited for what we have in mind."

"And semi-literate with an above average IQ, you say?"

"Yes."

"Apt for our mission, I guess."

"Definitely - speaks the local lingo, along with a smattering of the English language as well."

"Good! A perfect specimen. How long before he revives from this state of oblivion?"

"Not until he is transported back to where he belongs; there'll be an intermission of 3 to 4 Martian hours when he will be ushered out of his stupor to assess the success of the implant."

"Is that a given?" asked Maxus disbelievingly.

"Virtually. The computation on which the armature of the entire procedure has been conceived is watertight and utterly unswerving, leaving no room for doubts on that count."

"Very interesting!"

"And more importantly, credible! Incredibly credible!"

"Excellent!"

"It's imperative to start instantly," expressed Pontus, his tone betraying the urgency of the situation. "Immediacy is obligatory to the assignment, and a prerequisite to building the right momentum."

"Signal the 'Intelligentsia', at once," instructed Maxus.

The 'Intelligentsia' was a synergy of Martians whose superiority of talent, intelligence and resourcefulness surpassed the rest. Chalking out lines of attack to help establish an unchallenged supremacy in Space, was one of their myriad responsibilities.

Sworn to secrecy, these highly competent intellectuals constantly polished their knowledge and skills by keeping themselves abreast with the latest affairs in the Universe. Their instinctive capacities to deal with great quantities helped them closely monitor detalles tecnicos with a degree of expeditiousness appurtenant to the assignment in hand. Strategies were subjected to thorough analysis before actual execution to guarantee the success envisioned by them. Bottlenecks were honed to perfection and reflections were pushed to accomplish an unflinching unassailable, single goal - to rule the Galaxy.

The lesser Martians only followed orders.

* * *

Gattu was piloted into a transpicuous bubble shaped 'Max Mars Lab' and mounted on a raised platform temporarily designed for the purpose. Queerly oval, the berth had an ignis fatuus with an elusive cerise tint all around it. Fierce dominant rufescent flames interspaced by a subdued indistinguishable bluish tinge leapt up in spurts sending sparks into the atmosphere at intervals. It vibrated for a few seconds before becoming steady, after Gattu was positioned.

Pontus introduced the visitant and announced in unequivocal terms, "The Earthlings have accelerated their momentum. Their cosmic surveys, their nosiness, their interminable assay and periodic haunts to our planet have become increasingly threatening. So, before any of their actions constrain us to subservience, we need to secure ourselves."

A pall of gloom descended on the congregation. That these Earthlings would prove to be far superior to them, in the foreseen future of the Cosmos, was bad news! They had to be defeated with a vengeance.

"'Tis a pity the bounteousness of Earth stands intermingled with pestiferous human forms. Life would have been a lot easier without them. It would do us no good to be mired in a situation that alienates us from our ideologies," warned Pontus.

The silence that followed this pronouncement underscored the sobriety of the predicament.

"But now," continued Pontus, "we have an opportunity to realign ourselves and negate all errors of the past. We can't afford a single mistake, a single solecism. Jockeying this proposition to success is supreme. Remember, the supremacy of Mars - its very dominance in the Cosmos - is at stake."

In the Martian territory no elaborate discussions were ever necessary. Conversations were restricted to salient questions and doable solutions. Decisions were wrapped up swiftly without much ado. The gravitas of their current situation loomed large in their minds. They were aware that their capitulations were necessary to confer a quality of irrevocability to the 'Plan.' Striking the right balance was imperative. It would be injudicious to rework the same merely for want of an insignificant detail either due to carelessness or due to overconfidence.

Martian 'A' was the first to respond, "The Master Plan? Has it been decided upon?"

"Yes," replied Pontus emphatically.

"Alpha or Gamma?"

"Gamma," accentuated Pontus.

'Gamma' was a reference to programmes that were notoriously surreptitious in implementation. They were slow in giving results but only initially. When they did pick up pace, they did so by leaps and bounds leaving little or no room for interlopers to react, unless they were microscopically nimble. Their execution was so subtle that the prospects of a whip-smart intrusion were exceedingly remote; but, if intercepted by a twist of circumstances, they got circumvented at the preliminary stage itself.

In this case, the single element on which everything hinged was the placement of the microchip. It had to have the element of adequacy to avoid discomfort to the carrier, but more importantly it had to be so positioned as to produce the desired results for which it was configured. It was so deceptively simple that the Earthlings who sought snags even in the simplest, would overlook it in sheer disbelief, smug in the thought that it was unlikely that something like this would actually work at all, or so Pontus thought. And yet it would–with shattering, devastating consequences.

The dazzling red flames hemming the make shift dais on which Gattu lay, continued to shoot up brilliantly as Pontus surveyed, with pride, his battery of accomplished individuals. He was confident that the newest proposal would be as effective a triumph as he had visualised.

"The microchip?" asked Martian A. "Where do we sneak it? The heart or the head?"

Pontus let his eyes wander, resting them briefly on each of the Martians present. He was seeking consensus while simultaneously weighing the alternatives. "Heart or Head?" he posed.

"Both the heart and the head," riposted Martian B.

"Two chips? We need to employ just one. What makes you advocate something so obtuse?" rebuked Pontus.

"To avoid a glitch," replied Martian B. "If one collapses, the other will function."

Pontus glared. It was obvious that the commendation was repugnant to him. One did not contemplate failure even before the Project was implemented. 'Twas anathema. He darted a quick glance at the others, emboldening them to voice their opinion.

It heartened him to note that Martian C disagreed.

"Not the wisest of things to do," said Martian C. "Reports suggest that self-discipline is a habit that's hard for the Earthlings to inculcate. They are a confused lot. When confronted by ambiguity, they keep dithering between being cold-bloodedly practical and emotionally foolish. The continuous toss between the heart and the head may not give the coveted results. Precisely why targeting both may not be prudent."

The opinion was underscored with deliberate harshness.

Pontus gazed at him appreciatively. "Logical. Do we then set our sights on the heart?"

It was not a statement but a question that sought an answer.

"That wouldn't be prudent either," said Martian D.

"Why not?" asked Pontus.

"The human heart is hyperactive," explained Martian D. "A slightest shift in emotions leads to a differential in the heart beats. Counts increase when stimulated and decrease when subdued. Then there are instants when the body is stilled and the heart stops palpitating in a jiffy causing immobility of speech and action. This swing from one extreme to another interspersed with a hiatus may cause havoc. It could be detrimental to our 'Plan' which entails a need for permanence."

His tone accentuated conviction. He paused to ascertain whether his argument had carried weight. Apparently, it had. There were nods of approval all around.

"An excellent observation," responded Pontus leading him to make an obvious decision. "The brain it should be then. It is the most intellectual with an ability to function exactly like our robots here. Once pre-recorded, it gears itself to follow instructions staunchly most of the time. Of course there are moments when the haughtier among them make an arduous effort to appear far more intelligent and wiser than they are. These arrogant individuals, then attempt to introduce some superficial doctrine which they hope will influence their lesser brethren to behave in a contrived fashion. But such moments are far and few."

"I don't understand," murmured Martian B.

Pontus drifted from the task fleetingly to elaborate, "Well! the human brain, though primitive, is both smart and dumb. Its smartness lies in its ability to decipher, disseminate, choose and pick, though not necessarily sensibly, and its stupidity lies in its weakness to be influenced easily. There are all kinds of humans down there, but the worst are those that connive. These 'scheming' beings are aware of the duality of the brain, and use it to suit their ambitions. They impose their views and force the more innocent ones into thinking that their brain is far superior to the rest. Their Ego then takes over. Result - intolerance, strife, uproar, war."

"Ego?" Martian B looked at him cynically.

"Yes Ego! A human being's biggest failing! The egocentric individuals are so guided by their ego that once they compel others to believe that they are a cut above the others, they become dictatorial. They then obsessively hunt for weak targets suffering from a herd mentality. If they succeed in achieving their goal of entrapping powerless beings, their complacency leads them to presume that they have been victorious in enslaving a following that apes them in thought, word and deed. An individual that houses such an egoistic brain then moves around arrogantly with a title of a 'Hero' or a 'Leader,' but not for long. The more ambitious ones refuse to be subjugated. They try to supersede him by conniving in his downfall. So much for the predictability of the human collective nature!"

"But of what value is that to us?" countered Martian B.

"Of great value!" Pontus asserted. "I am trying to explain, in a reasonably straightforward manner, how central an organ their brain is. We control it; we manipulate the person who houses it. Do you understand?"

The Martians agreed raucously.

Pontus added, "Besides, the innumerable fissures of the brain are an added bonus. The size of the chip coupled with the knotty structure of the brain will let the chip rest there undiscovered forever. By housing the microchip in this cerebrum tank, we can rest assured that it will play a dual role; it will not just do what is intended of it, but also lie innocently ensconced from the prying human eye."

The rationalization satisfied the Martians. There were only thoughtful faces telling Pontus silently that his ratiocination had struck home.

Pontus then removed an imperceptible tablet from behind his ears and tapped out directives with a slim electronic pin. Instructions flashed across a CRT screen placed in the line of vision of all the Martians aggregated for the purpose.

"The text. It's here. It has to be followed to the letter," Pontus said with an easy assurance, "it will help you skilfully position the microchip without major manoeuvring."

The Martians expressed their understanding with an equally easy composure.

Pontus paused perceptibly before looking at his squad searchingly, "And now for the most important question; how long before the entire surgical operation is concluded?"

Martian brains were astute enough to perform competencies where error and correction were not acceptable. Calculations were unvaryingly worked out in Martian milliseconds with adequate latitude for improvisations, if needed. So, when Pontus looked at him categorically, Martian A calculated swiftly, "About 3–4 Martian hours assuming everything moves as detailed. This, of course, omits the additional Martian minutes that would be needed to test the success of the implant."

Pontus acknowledged his appreciation, "Good. We better be quick. As crucial as the surgery is, equally imperative is to situate him in the original spot from where he was hauled, at the earliest. We want no theories or hypotheses on his sudden evaporation to be parried and mired by snafus of heterogeneous human deliberations. Until sometime ago, I would have assertively stated that the likelihood of his disappearance being traced to us is extremely remote. Not anymore; the rapidity with which the human brain is evolving is alarming. Let's begin."

Pontus lingered for a split Martian second before continuing, "One other thing, don't forget to keep pumping him with oxygen at proper intervals. Remember, a human being needs oxygen to survive all the time, and every time. We want no carelessness from our end to be responsible for his death.

"How intricate! Is that how they subsist on Earth?" asked Martian 'C' curiously.

"Is that how...?" trailed Pontus looking at him irritably.

"I mean, do they have to constantly infuse themselves with oxygen all the time, just to stay alive?"

Pontus laughed. He found the question rather flippant and humorous. "No, no. The atmosphere down there is plenteously laden with it. The human body is acclimatized to keep inhaling and exhaling it alternately through its nostrils from the time it is born."

"Then surviving and living in such climes, is presumably not as byzantine, as I had thought," said Martian C.

It was a statement that anticipated no answer, but Pontus heard it all the same and proceeded to explain, "It is not, but they have convoluted their life by cosmetic inventions. Don't forget that oxygen is just one of the many parameters on which their life fulcrums, but uncontestably the most important; a mere 3 to 4 Earth minutes without it, can cause irreparable damage to their brain."

"And the others?" Martian C queried inquisitively.

"Others? There's water, and something else that they refer to as 'food' which is sometimes heated and at other times not. Oxygen and water are just simply available in their planet while 'food' is far more labyrinthine. There's plentiful of that too but, as with everything else that is abundantly available there, they have created a complication out of it by experimenting with it."

"Experimenting with food?" asked Martian C

"Yes. You've heard me right," declared Pontus. "They have begun subjecting it to a process called 'cooking.' This 'cooked food' needs to be well ingested to provide the necessary nutrients to their skeletal frame with its cushioned covering of flesh and skin. A system called the digestive system breaks down the consumed food within the human body, and a part of the assimilated nourishment is absorbed by the blood stream while the rest is excreted as faecal matter. The Red Blood Corpuscles, one of the more important components of their blood, play a vital role in promoting the oxygen levels in their body."

"A difficult life indeed!" This time it was Martian 'A' who responded.

"Yes! An exigent life indeed, made exigent by the humans themselves. They love intricacies. It helps the degenerate to assert prepollency over the guileless," echoed Pontus. "It's no wonder then, that they are so impuissant to death and lead such remarkably condensed lives."

"Then why do we have to wipe them out?" asked Martian C. "They'll destroy themselves."

"It's their propensity to multiply faster than the rate at which they perish that is a cause for alarm."

Martian B looked at Pontus with admiration. "Fascinating! Your store of facts is admirable. The research on the subject is nothing less than top notch."

"Correction! Not mine, ours. These established inferences are the outcome of the research of our entire consociation which is continually keeping itself abreast of the latest intelligence of the other entities in Space. It is imperative, for us, to fill our coffers with more and more knowledge to ensure that no stone is left unturned in our mission to dominate the Galaxy someday."

Pontus paused. He wanted to say more but then decided against it.

"This is not the time for a symposium on the subject," he said, putting an end to the topic under discussion. "Our priority, now, is to complete the task in hand without enkindling human intellection, and more importantly without arousing the invasiveness of those snooping Venusians. They need to be shown their rightful place in Space. Their incessant prying and interference can cause untold damage. It's time they realised who actually rules the Cosmos."

"Agreed!" chorused the Martians in unison.

* * *

The mood in the Martian domain was celebratory. The implant was a theoretical success. It had yet to be tested practically, but Pontus judged that there would be no setback on that front. It was a fundamental principle of his professionalism to avoid implementing a project unless it was well stripped of likely gaffes. But his past experiences had taught him to be wary of overconfidence. There always was a possibility, remote as it seemed, that there may have been an accidental goof-up. The only dynamic that would decide if their present accomplishment was indeed 'unblemished' was the 'outcome.'

The 'Maxx Martian Auditorium' was readied for a live presentation to facilitate the Martians to have a peek at the pantomime as it unfolded itself and offer their critique or appreciation, if any. He hoped there was no disappointment.

"Pantomime?" questioned Maxus

"It'll be no less an experience," responded Pontus. "Your interpretation would be critical to its psychology."

"When will he be brought in?" asked Maxus whose rigorously precise intellect always kept him in a state of circumspection. If this did not give the anticipated outcome, it would necessitate a tearing down of the entire project and rebuilding it. As he became absorbed in the various twists and possibilities, Pontus put a brake to his flow of thoughts.

"Immediately," Pontus said. "But it will be a while before he awakens to an induced fugacious sentience, and gets animated by the presence of 4 robots placed directly in the line of his vision - 2 human males and 2 females. His reaction will decide our triumph."

"What exactly is expected of me?" asked Maxus

"Your observations, your comments, your impressions; nothing else," replied Pontus crisply.

Maxus expressed his understanding with a slight nod of his head, and watched the ongoings with dispassionate objectivity as Gattu was negotiated in.

After a brief period of nothingness Gattu appeared to be responding; his limbs moved, and his eyelids flickered. He sat upright for a while and then lapsed into a drowsy stupor. In another few moments he would have relapsed into a comatose state but for the agility of Pontus who was quick to notice what others had not.

The air here was rarefied for a human being, and unless a fresh spurt of oxygen was pumped in, it wouldn't be long before this creature would die. Immense precautions had been taken to ensure a steady flow of oxygen in the surgery blimp but the same had been condoned here. Pontus felt feckless and blamed himself for overlooking this all important factor. His incompetence would have galvanized yet another catastrophe. Amends had to be made before everything dissolved into zilch.

"Quick," Pontus commanded, "impel oxygen from the chamber through the gaping hole at the rear end of the amphitheatre. The constitution in the atmosphere needs to be restored to the echelon suitable for this Earthling within the next 5 Martian seconds to stop him from lapsing into total unconsciousness."

A bustle of hectic activity followed. Soon everything held together.

Gattu sluggishly opened his eyes wondering at the absolute blackness that inundated him. It was baffling. There was no light - only darkness, and yet he could see unfathomable mavericks. He failed to comprehend his ability to discern figures despite the lack of light. Or was there light? Was it blurred or opaque?

In his state of bewilderment, Gattu was unable to achieve clarity in his thoughts. There was an astonishing shade of red that appeared to be leaping like flames out of nowhere. It was uncharted territory for him; unexplored, unmapped and frightening. And what or who were these critters that flitted around him? They certainly did not belong to his world.

He struggled to get up from his horizontal position. It was tough going because his legs were all but useless. He gripped hard on to the inflexible alloy that apparently held him. His hands lost their grip and he was obliged to clamp the composite with both his hands and feet. Feeling breathless he thought it wise to lie in that stance for a while. When he sensed that his lower limbs had returned to a semblance of normalcy, he heaved himself upright. He felt light - a sensation of floating engulfed him.

Straight ahead he saw 4 human beings–2 males and 2 females, but they seemed unreal, as if belonging to another era. He gave a loud involuntary scream—a scream that stemmed from a fear that unambiguously disclaimed any personal knowledge of the environment he found himself in.

"What's he doing? Stop him from making such a weird sound," yelled Maxus.

"Ignore his reaction," conciliated Pontus. "We cannot let ourselves be distracted by such minor irritants. He'll slow down. It's a perfunctory reaction of all human beings in a foreign setting. There are far too many particularities for him to grapple with. He senses danger, and hence the corollary. Once we have adjudged triumph of our experimentation he will be anaesthetized to preclude him from grasping the rush of events swirling around him. Just concentrate on his eyes."

Maxus looked punctiliously, paying attention for the unusual.

"See! His eyes are flickering," exclaimed Maxus.

"That's great! He'll react sooner than we anticipated," replied Pontus.

Maxus stared intently at Gattu's eyes and continued with his commentary. "He has focussed on the male model. He is blinking; 3 times in quick succession. Now what's that?... A faint reddish light emitting from his eyes... It's advancing to the eyes of the male model."

"Just quite what it should be," stated Pontus.

"But there's no reaction," said Maxus.

"Predictably so," replied Pontus. "No reaction, none whatsoever. Now, observe intently what happens when he focuses on the female."

"He is blinking; 3 times in quick succession again," commented Maxus

"Is that all?" asked Pontus

"A similar faint red light has effused from his eyes and as in the earlier case, it is proceeding towards her eyes. Amazing! She's reacting. She's inclining. The stoop is barely discernible," reacted a visibly excited Maxus.

"Precisely," said Pontus crisply. "This implies that he has accomplished what he has been attuned to do. He has transferred the imprint of the chip into her brain. The red rays play a dual role; that of a vehicle and a catalyst."

"An irreproachable transaction!" commended Maxus, "but not necessarily a guaranteed one. Don't misconstrue my Pyrrhonism, but it would be foolhardy to underrate our enemy. Odds are that they may notice what I just did. What then?"

Pontus grimaced, "They never will, unless they are exceptionally quick. They deal exclusively in the present and lack the patience to see beyond a horizon they may never see. And yet, for the sake of an argument, let us presume that you are correct in your postulation and they do notice it, then their actions will be restricted to just their observation and nothing beyond."

"That's encouraging, though not entirely convincing," said Maxus with a hint of uncertainty in his voice. "Anyway, apprehensions aside, what happens once the chip is embedded in the human brain?"

"It stays there and functions the way it is meant to. An encounter with a female will instigate a reaction from her - the kind you noticed in the female robot over here. Death will occur after multiple encounters. 'Multiple encounters' are deliberate, to avoid suspicion, detection and alertness," conveyed Pontus.

"Ummmmm..." Maxus mumbled paying rapt attention to Pontus.

"Every sequential gaze from the source, will make the target feebler and weaker, crippling it to such an extent that its normal body functions will cease to perform efficiently. No doctor, however brilliant, will be able to diagnose the cause. For your information, a doctor is a medical practitioner who is qualified to treat ill patients on Earth," articulated Pontus.

* * *

CHAPTER III

The perspective given by Pontus had one big perquisite. It offered anonymity - absolute anonymity. Why would anyone on Earth correlate death of females to a planet Mars.

"So, this is the modus operandi... annihilating females slowly but surely," said Maxus.

"Yes! The first part; to eradicate all females," replied Pontus.

"Eradicate all females?" Maxus sounded surprised. "I thought our schema was configured to decrease the number of females in comparison to the number of males."

"The initial blueprint had entailed just that," clarified Pontus, "but as we went over the basic arithmetic there was a need to recheck the original criterion. The concept of a skewed ratio of the male and female was premeditated from a need to avoid suspicion from the meddlesome Venusians who are desirous of a ratio that tilts favourably towards the females."

Maxus put a brake to his reasoning, "Let me remind you, without seeming too presumptuous, that the same officious Venusians were responsible for the premature abortion of our last venture. We can't overlook that.

It's paramount to factor in their interference to experience the kind of success we fancy. Like it or not, they continue to hang in the shadows. There's every possibility that their suspicions will always be targeted at us irrespective of whether we are responsible for female deaths or not."

"I have not forgotten that," professed Pontus. "How would I? You never let me do so. You reminded me of them when I just entered our territory with the human sample, and you remind me now and you will always keep reminding me of them. I don't deny that your reminders are necessary, but when they start becoming an obsession, we'll have to let go of our goal. We can't let our phobia for them throw a spanner in our works, all the time."

Maxus looked at Pontus looked rather appallingly. He thought Pontus sounded reckless.

Pontus justified his words with a more expanded elucidation, "Nothing risked, nothing gained. It's time we stopped treating the Venusians as a powerful deterrent. It is true we need to feature their infringement in all our game plans, but it is far more crucial to work on programmes that give deliverables to suit our ideologies. Their constant attention is an unavoidable irritant that must be dealt with adroitly."

"Precisely! That's why the need for caution," reproved Maxus.

"But elimination of the entire female species from the face of Earth is peremptory," argued Pontus. "It's a crazy world out there and even a single surviving female would be a threat. As long as they exist, the probability of stopping the Earth from its paralysing monotony of giving birth to more such females would be very low."

"I got it," acquiesced Maxus in a placatory manner. It would do no good to discourage Pontus, not when he was all geared for the Project in hand. "And what about the second part?"

"The second part needs no involvement from our end. No females, no procreation; the males will wither, and the human species wiped out, leaving the planet pregnable. That's where we enter to make a rightful claim," justified Pontus

"Sheer brilliance!" commended Maxus. "But slow deaths aren't my cup of tea. Can't we just get them killed straight away to empower proprietary over the Earth at the soonest?"

"Not at the cost of drawing undue attention of the Venusians," advised Pontus.

"A contradiction!" mocked Maxus. "You did comment a while ago that their 'attention' is not a constraint."

Pontus averred, "Let me put that in the right perspective. I mentioned that their 'attention' should not be a deterrent. I did not say that their 'attention' should be totally disregarded. They possess an irreproachable data on the Cosmos. Their record of the male:female ratio on Earth is impeccable. Even the slightest imbalance will arouse their suspicions and as you rightly pointed out, irrespective of who is responsible, we'll always be their key suspects. I have not and never will forget how they foiled our plans even before take off the last time."

"Let me not be reminded of that," Maxus stated. "And let me also remind you that it is you who speak of it now, not I."

"The occasion warrants it," declared Pontus

"Does it?" ridiculed Maxus. "But while at it, I would like to extend the subject further. It's been on my mind for quite some time and I'd appreciate your assistance in helping me clear my perplexity."

Pontus cocked his head in a show of attention.

Maxus proceeded, "I am still undecided by our attitude to their mediation. Were we humbled by them or humiliated?"

"Both," affirmed Pontus, "Why do you think we have to factor them, every time we devise a ploy? Why do you think you have to constantly keep reminding me of them? Because they have a displeasing tendency to interfere when we least expect them to. They are most certainly not Herculean, but their constant intervention does make us appear superfluous in the part of the Universe where we are expected to dominate. That is why it would be imprudent to disregard their 'attention' but contradictorily we cannot let this 'attention' be a disincentive. Paradoxical, isn't it?"

"It is," said Maxus emphatically.

"Our present Plan," announced Pontus, "is just that - a paradox. It has both 'distraction' and 'action.' Their distraction towards anything but Earth, and our action towards nothing but Earth. This slow machination will lull them into complacency while we carry on with our démarche. Quick deaths will arouse the referenced 'attention' that we are judiciously trying to avoid."

Maxus admired his ingenuity. "That's a precocious approach."

A long silence ensued. Maxus was the first to speak. "To me, the attempts of the Venusians to redress the ostensible injustice done by us have always seemed phoney. I have never really comprehended their penchant for that planet."

Pontus shook his head dispiritedly, "Neither have I. It probably stems from their strong creed that if a ratio of 'females outnumbering males' is not compulsorily upheld, 'Space' would witness colossal disaster, and the Solar System that we belong to, will dissolve into a Black Hole. It's a theory yet to be proved; yet they continue to cling on to it as if it's an accepted diktat of the Universe."

"What prompts such a hypothesis, I wonder!" contemplated Maxus loudly.

"Their lopsided assessment that females are harbingers of peace while males symbolise violence," said Pontus sarcastically.

"Just like females to override logic for emotion," snorted Maxus.

"Agreed," acceded Pontus. "They may have a spectacular capacity to compress lifetimes into a matter of moments but when it comes to logic which is so manageable, their attitude deserves nothing less than contempt. Besides, I fear they may have some other motives too."

Maxus straightened up, "Other motives?'

"Yes," stated Pontus prosaically. "Just a stray thought but certainly not beyond the realm of possibility. The theory may be off-centred, and I hope I'm not stretching it too far when I state that they fear an invasion of their own planet by us, the moment Earth becomes a prized heritage of the Martian map."

"Are they justified in thinking so?" asked Maxus.

"Perhaps, yes," professed Pontus. "Look at it from their standpoint. It would be a simple enough matter for us to embark on an invasion of Venus, using Earth as a base, since its distance to Venus is easier negotiated than from Mars."

"Ah! Self interest? I guess that's what actually underscores their vigilance on all our affairs concerning Earth, and not necessarily protection of Earth itself," said Maxus. "It's no wonder then, that their moves to subjugate us are designed with such tactical rapidity, but your guesstimate does deserve consideration. It would not be a bad idea, after all, to annexe yet another planet, but their simmering temperature is an obstacle. A great obstacle."

Pontus concurred, "I'm on the same wavelength with you. I refer to the annexation. It's a possibility that's feasible. As for the heat, time has a way of finding solutions to all problems, no matter how ginormous."

Maxus reflected, "We'll have a discussion on this at leisure. Right now, all I can say is that I'm beginning to understand the wisdom behind this unhurried ruse. It makes better sense to me now."

"Yes!" granted Pontus. "A subterfuge that is confined to the highway, compelling the Venusians to be unaware of the dark possibility of chicanery. Females will die but the deaths will be a secret to the Venusians and an enigma to the Earthlings. When the realisation dawns that something untoward is occurring, it will be too late. You know what the Earthlings say, 'Slow and steady wins the race.' We have to win the race this time with this slow action."

"Wow!" Maxus exclaimed with awe.

A weighty stillness suffused the atmosphere. Maxus diluted it with a rational question, "When do we haul this empirical object back to Earth?"

"As soon as we are done with a microsurgical cut," said Pontus pointedly.

"Is that sine qua non?"

"Most certainly," confirmed Pontus. "Remember, the basic precept of all our missions is to bereave them of all telltale signs. To that effect, vaporization of memories of his stay here is mandatory. His brain will have to be attuned to recall just his last moment on Earth before we picked him up. He will be positioned at the exact spot soon after the 'memory swipe' operation. It is imperative that he is discovered within an hour of his deposit which on rough calculation would be around 19.30 Earth hours. Any delay may result in him being undetected for a good 7–8 Earth hours, and that may be calamitous."

"Calamitous?"

"Yes! The timing is deliberate. Neither day nor night on Earth; just when the solar body is about to set there, leading to a graceful dissolution of day into night. There has to be enough light for him to be discovered. A little late, then darkness would descend. The location he's to be consigned to is full of prowling vicious beasts. Even a slight deferral may result in a feast for those wild animals which show a fondness for being active at a time when humans lull themselves into deep slumber, blissfully unaware of what's happening to one of their kin."

"Slumber? Now what is that?"

"The function is referred to as 'sleep.' They close their eyes, and simultaneously put down the shutters of their conscious mind. In that state, they are, by and large, unaware of what happens to the world around them."

"How amusing! They seem to be doing a lot of things... fight... cook... eat... sleep... it's no wonder then that they are born with such a short expiry date."

***

The wee hours of the morning had heard faint rumblings of a thunderstorm with stratocumulus, fleeting, fast moving clouds hovering over the place and threatening a ruthless downpour. But the clouds had chosen to sail past leaving behind a blue sky and a weather so fine, that it needed to be venerated.

It was almost sunset when Dondu entered the jungles. He was on an errand to deliver an important message to Chachu, a responsibility characteristically shouldered by Gattu. But with Gattu's precipitous disappearance, the onus had fallen on him. The others had declined. They were afraid of cutting across the wild wilderness that was perceived as being unsafe, at that hour. It was late, but not too late. He looked up to see a rare sight; a vision of both the Sun and the Moon—one setting and the other rising.

The cool breeze caressed his robust cheeks as he jaunted leisurely down the pathway enjoying every moment of his stroll. Like most others, he too generally avoided taking the short route, but today he did so, because his instinct dictated otherwise. He followed what he sensed was probably a hunting trail that was far narrower and rougher than the one on the longer track. At specific patches, the whipped branches of the undomesticated trees obscured his vision.

Letting his eyes wander on the changing panorama, he halted on and off to admire the splendours of nature which were lacking on the lengthier route. Abruptly he stopped. A peculiar red glow at a distance, rising out of nowhere, attracted his attention. It was eldritch, spine-tingling and did not quite fit into the jungle scene. There was a strange buzz too; like a swarm of honey bees being hounded out of their honeycomb and preparing to sting anyone who came their way.

His phlegmatic stride decelerated to a reluctant walk. He had another peek in the direction to reconfirm what he had seen. Yes, it was unmistakeably there. Fear paralysed him transitorily. He dithered before turning and sprinting through the unruly, untamed footpath back to the Master's house, as swiftly as his feet could carry him. Incoherent questions kept cruising through his mind... Spirits? Ghosts? Who were they? Or what was it? He cursed himself for having let his gut feeling predominate his choice for the shorter way. "Damn it! Never again," he stuttered to himself, as he kept giving surreptitious, quick glances behind him, every now and then, to ensure that no inhuman stalked him.

"Master! Master!" He panted.

The Master, who had just temporarily retired to a contemplative evening after a light supper, hurriedly got up when he saw Dondu rushing in puffing.

"What's wrong? Why are you so disturbed?"

"There's a queer red glow in the forest," Dondu said breathlessly.

"Relax! Relax! Get your breath back," said the Master.

"Red... red..." Dondu stammered. "There's a supernatural red glow... in the woods..."

"A red glow? Probably cast by the setting Sun."

"No... it's a red glow that's whacky!"

"A red glow that's whacky? Hallucinating? Are you?"

Dondu gave a soundless laugh, "No, I'm certain. I saw it. It was eerie. There are ghosts revelling there. I'm sure. I swear... its true... I saw... I..."

His shambolic words intermittently punctuated by a series of huffs, puffs and pants took a good 5 minutes before they subsided into a sotto voce tone. The Master was at his wits end. To believe or not was the question, but there was something almost fanatical in his tonality. He let him unwind for a few seconds during which the clock on the wall ticked chillingly.

"Calm down, Dondu. You are probably tired. Rest a while. I'll ask Ramu to give you a glass of water," the Master said in a subdued authoritarian voice inflecting his phrases with an underlining reprimand.

Dondu shook his head, but he wasn't listening. His mind was reliving the haunting moment he had encountered in the jungle. He gulped down the water offered to him and stared blankly into space. The beads of perspiration that had formed strings on his forehead kept falling drop by drop down on to his cheeks, and then on to his neck. He was nervous and kept biting his nails furiously in a bid to hide his jitteriness.

"Now you can try telling me again," the Master goaded.

Dondu moistened his dried lips with his tongue and attempted to impart a degree of coherence to his narration. "I was going to Chachu's house to convey your message," he said in a voice that guised itself in an agitated whisper. "I took the shorter sentier just so that I could reach his house faster, when I noticed this red glow - luminous, foreign and indubitably not a part of the landscape. I should know. I was born here, I belong here, and I recognise every little thing that belongs to this place like the back of my hand. And then there was an uncanny sound that kept vibrating in my ear—as if a thousand bees were swarming there. I was on pins and needles and wanted to investigate but had no gumption."

The Master was undecided on his reaction. He was familiar with Dondu's histrionics. His words were always taken with a pinch of salt, prima facie, as he loved to exaggerate even the most insignificant. He gave him an indefinable look, and then another, to gauge the seriousness of his prattle. He certainly didn't appear insouciant and for that matter, his tone didn't sound facetious either. The Master decided to give him the benefit of doubt but not before putting his mind to rest by asking one final question to quell his suspicions.

"Are you telling the truth Dondu, or is it yet another of your attention seeking gimmicks? Tell me honestly," he asked sounding a little cross.

"I swear I'm telling the truth," Dondu persisted. His perturbance had to be seen to be believed. He wished he had not cried 'wolf' in the past.

The Master hesitated before relenting, "In that case, we better summon the rest of the men in the house and find out. Should nothing of consequence be found, I will sack you for wasting our time." He spoke with a mordant asperity.

An hour later, a group of motley men armed with sickles, sticks, clubs and the Master himself with his rifle went to the woods, and guided by Dondu, cautiously moved into the jungles and towards the spot indicated by him, to intercept the monstrous will o' the wisp. As they converged onto the area, a sky-cleaving streak of lightning followed immediately by the rumblings of thunder threatened to change the calm weather into an ill-omened one. All eyes kept searching but no quirky lucency was visible.

Dondu was perplexed. He was sure as eggs is eggs that it had loomed large, only a few instants earlier, while he had crisscrossed through the gnarled roots of the thick set trees with knotted branches. His credibility, which was never worth a grain of salt, was at stake. It was incumbent on him to prove that he had not narrated a fanciful story; but with no concrete evidence, he felt trapped by his inability to convince these people. He walked ahead of the gang until he thought he had reached the exact section of the forest where he had noticed it. There was nothing. The crimson irradiance had apparently dissolved, leaving behind nothing but the colours of the night.

Tugging his hair in utter frustration, he shook his head woefully and deliberated for a moment. He knew he had not been erroneous. With a forlorn expression, he mentally prepared himself for ridicule. He turned and took a few short steps to return to the spot where the others waited for him, when he espied a body sprawled a few metres away from where he stood. Cutting diagonally across, he tottered towards it. It was Gattu.

He called out excitedly. "Look! Gattu! He's here. Come fast. How the hell did he land here?"

One of the other servants, Ramu, came running to check if Dondu was telling the truth or pulling yet another fast one on them. Confounded by the presence of the body, he resonated what Dondu had said earlier, exclaiming loudly, "Why! It's Gattu!"

The clique that surrounded his body was hysterical. Wasn't it just a few days ago, that they had combed this very thicket meticulously? He'd been untraceable then. But now he was found in the very place that had not the slightest vestige of a human form. The mystery was inexplicable. The crowd inched closer.

"Yes, it's Gattu!" bawled all of them together in unified affirmation.

"Are you sure?" bellowed the Master who was still some distance away from the scene. "It's hardly been a while since we scanned the entire jungle. He was nowhere to be found then, and now you say that he is lying exactly where he never was. Come on, let me have a look." He approximated to the spot where he lay and checked his heartbeats by placing his ear against his chest. "He's alive," he opined, "but unconscious.

It's a miracle and a peculiar one at that. Get to work all of you. Pick him up and take him to the house. He needs to be attended to by a doctor."

At the Master's house, Gattu was placed on a soft rug that served as an impromptu carpet. The only doctor on call, lived about a mile away. When Ramu reached the general practitioner's clinic, he was informed that the physician was attending a medical conference abroad, and was not expected to return until after a week. Seven days was a long time. The Master was left with no choice but to wait for Gattu to regain consciousness on his own.

After a seemingly long delay, almost 17 to 18 hours later, Gattu slowly opened his eyes. "Where... where... where... am I?" He stammered in a whisper that sounded incomprehensible to him against the insistent ringing in his ears.

"You are at my house," said the Master. "How are you feeling now? And what happened?"

Gattu looked at the Master sketchily before flashing his eyes in a sign of recognition. He tried heaving himself to a standing position, but unfathomable physical restraints prevented him from doing so. He felt unbelievably heavier than usual; his mouth felt dry and a peculiar metallic smell in his nostrils made him nauseous. Uneasily conscious, he pondered at the surroundings in which he found himself.

"How come I am here? What happened?" he stuttered making a futile attempt to hide his embarrassment. He felt like a fish out of water despite being in a familiar environment. Oddly he also felt a little bolder, less submissive. He wondered what the commotion was all about.

"Exactly the question we want to ask of you. What happened? And, where were you?" said the Master. "For the past few days we have been scouring places in and around the jungle to locate you, but you were untraceable... that is, until now. It's a great stroke of luck that Dondu stumbled upon you in an odd location in the jungle. Do you remember anything?"

Amnesia gripped Gattu. His mind experienced a total blank; a blankness that slowly but surely began to be superseded by a few scattered but distinct images. Then, in a spurt, like an oncoming locomotive hurtling out of a dark tunnel, he vividly remembered going home from the Master's house whistling the English rhyme that the schoolmaster had recently taught him. He also recollected seeing a mesmerizing reddish tincture at a distance. "Much against my will, I found myself drawn towards this entrancing colour. Hypnotized, I inched closer and closer towards it. What happened after that, I do not know. And now I find myself here amidst all of you," Gattu concluded his narration in a deliberate manner as if attempting to reinforce his words with honesty.

"See, I told you," blurted Dondu. "I told you that I did see a queer red light. If Gattu had also seen it, then I am certain that my mind was not playing tricks at that time."

"It was not you that I doubted," replied the Master, "but the complexities of the jungle which may have blurred your vision. Later, we can investigate further on this purported light that both of you seem to have seen, but right now the emphasis is on Gattu. Thank God he is safe!" Diverting his attention to Gattu, after that momentary distraction towards Dondu, he persuaded him to tax his brain further. "Try to remember, Gattu. There has to be an admissible explanation for your absence."

Gattu strained his memory, but it glued itself to the moment when he had last seen the bizarre red light in a forbidden territory of the forest.

"I can't," he said overexerting his mind as far as he could, just so that he could say something that would satisfy the people around him.

"It's okay. Relax," placated the Master giving him a fatherly pat on his head. "That you are safe is of greater importance than anything else. We can discuss this when you are in a better frame of mind. Right now, it's important that you reach home safe and sound. Your mother will be glad to see you. She has been anxious and worried. Besides, you need rest. Your eyes tell me so. They appear to be blinking abnormally."

* * *

As the twilight swept across the trees, the branches acquired a phantasmal aspect. A soft gentle breeze stirred the throngs of mosquitoes and forced the thick growth of poison ivy inhabiting the right earthen wall of her house to shiver spookily. Even the full moon failed to cast its light on the darkness that had descended upon her life. She had braved several storms in her life - some of them so severe that she had almost lost her will to live, but this one beat them all. It was the harshest. One by one, all her dearest possessions had been snatched away from her.

Nature had a knack of punctuating her life with sorrow every time she was on the verge of savouring contentment, without offering any pertinent reason. The last few days had seen her weeping profusely at her son's sudden disappearance. She wanted nothing more than to rewind the past so that she could feel his presence.

Running her fingers through her uncombed hair, she cast a cursory glance at the fire place that had remained unlit for the past few days. She hoped some genie would rise from there and grant her a wish. She would have asked for her son. Clutching his favourite shirt frenziedly she let her red swollen eyes meander across the horizon. A group of men at a distance apparently approaching her house, held her attention. As they grew closer and closer, she found herself staring disbelievingly at them.

"Gattu," she screamed fervently as she closed the distance with quick steps. She embraced him warmly, while all the time looking at him sceptically to ensure that this was not a dream. "Where was he?'" she enquired of Ramu, wiping her son's face haphazardly with the loose end of her carelessly draped saree.

"We found him unconscious in the forest, last evening," he replied and rambled on to narrate the details. "It took him a long time to regain consciousness. We were advised not to inform you until he had returned to a wakeful state. That is why we brought him now."

Pinching herself hard, she hugged him over and over again, and then ran to the disregarded hearth to rustle up a quick meal. Gattu was hungry and relished the rice porridge, so lovingly prepared by her. When his hunger had been satisfied he looked obliquely at his mother seeking some justification for all the commotion that hemmed him. She offered none. On the contrary, her attitude only served to increase the intensity of his discomfiture.

He wondered why his mother appeared to be harried, as she most obviously was. She was at sixes and sevens trying to validate what she thought she had just seen. Was her failing eyesight tricking her, or was she really seeing what she thought she saw? Yes, it was unmistakeable. The more she looked at him, the more she was convinced. His blinking! It was uncanny, it was supernatural. She wondered if the others had noticed it. Apparently they had not; they would have surely remarked about it, otherwise. The excitement of having found him had blinded them to the obvious.

"Is something wrong with your eyes?" she enquired.

"No," he replied.

"Are you sure?"

"Yes."

"Then why are you blinking like that?"

"Blinking? How?" he said in a tone of amazement.

"Almost as if you are trying to dislodge a foreign particle from your eye."

"Eh?"

"Let me get the mirror; you can see for yourself." The mirror was small and had been gifted to her by the Master's wife. She held it against his face and said, "Look."

His reflection in the mirror appalled him. He peered into it again, and then again. His mother was right. His blinking! It was indeed robotic, perfunctory and unnatural - there were 3 rapid blinks in quick succession, when there should have been just one. It had never been like this before. Through the miasma of self absorption that kept gathering around him, he had no viable explanation to offer. He experienced a whole gamut of emotions as he kept rubbernecking at the looking glass repeatedly. But both his body and mind were in no mood to solve the quandary. Right now, all he wanted to do was to sleep. Perhaps, as the Master said, he was tired, extremely tired. He heard his mother's voice, soft and gentle, "You are probably worn out after having gone missing. Try and get some sleep now. Maybe you will have something to say when you are refreshed."

After a fairly long catnap, Gattu woke up to the scarletish, orange colours of the sunrise and the morning tweets, chirps and coos. It was that part of the day when the mystic night receded gracefully to give way to the majestic appearance of the glorious Sun. The inscrutable headache that he had been forced to contend with the previous night was almost gone, but the feeling of awkwardness remained. "I'm hungry," he said.

His mother was relieved to hear his voice, but she failed not to notice, once again, the peculiar way in which he nictated 3 times consecutively in quick succession. She was angst-ridden. Soothing his head with a herbal balm that she had prepared while he was asleep, she queried softly in a voice that sounded more like a murmur, "I know I'm repeating myself, but are you sure you are not having any trouble with your eyes?"

"Honestly, no" Gattu replied. "I personally can't think of an explanation for this sudden blinking. I am bamboozled myself."

She nodded resignedly. If he said he was well, he probably was. "It's probably just the fatigue," was all that she said as she moved briskly towards the fire place to prepare a fresh meal of rice burgoo. Her brusque movement was slowed down by a sharp, sudden stab on her lower back. It struck her that she had felt a similar pain when she had looked at her son the previous evening; only this time it was more severe. Clasping the palm of her hand on to the area of pain and pressing it gently to relieve herself, she wondered at the oddity of the coincidence.

Despite being in her late thirties, her healthiness was quite robust. The harsh rural life had gifted her with an immunity that evaded most urban folks. Except for an occasional cold or cough her corporal well being had been exemplary. This twitch that she now felt, was abnormal. Continuing to compress her palm against the smarting spot, she walked resolutely to the fire place. She found herself stooping a little.

The hot rice gruel served with an equally hot spicy curry rejuvenated her son's spirits. He loved the way she prepared it. He felt as fit as a fiddle as he readied to present himself for work at the Master's house.

The monotony and dullness of the Master's house dramatically changed into a kaleidoscope of bustling activities. It was 'back to the grind.'

***

At 3 in the afternoon Gattu's mother encountered exhaustion. The pain at the lower back which was earlier endurable had worsened. Attributing it to the trauma of the past few days, she willed herself to relax. She rested for some time, and then geared herself to do her daily tasks of dusting the house, washing the clothes, collecting the fire wood and preparing the meals. It took her longer than usual to complete her household chores. She felt unusually dehydrated and weak; as if she had run a marathon. It never occurred to her that her listlessness was entirely at odds with the normal weariness she generally experienced at the end of each day. Gritting her teeth, she took deep breaths, resolutely ignoring the intermittent headache, but her head continued to pound. She washed her face and rinsed her mouth several times, in a bid to get rid of the enervation, but to no avail.

When Gattu returned home, he saw her lying on the rug in her usual corner. Her eyes were half closed and she looked pale. Her slightly greying hair lay damp against her cheek. This was so converse to her ordinarily belligerent attitude that it disquieted him.

"Anything wrong?" he asked.

"Just feeling a wee bit under the weather. Don't you worry; it's probably the climate and the recent carry ons. I think I'll retire to bed early tonight; need to compensate for the sleep that has evaded me for the past nights. I'll feel better in the morning."

He looked at her intently and touched her forehead to ascertain if she had fever. "No fever," he said with a worried expression, exposing his deep concern for her. It hassled him that despite the lack of apparent signs of fever or for that matter the lack of even a simple bout of cough and cold, she looked ashen.

Earlier in the day, at the Master's house, he had been bombarded with questions about his whereabouts during the 'alleged' missing phase of his life which had lingered so distinctly in the minds of those around him, but had for some strange reason absolutely stonewalled his memory. He had been determined to ask his mother about the veracity of those interrogations. But now, he was hesitant. He didn't want to worsen her well being by imperilling her with awkward quizzes. He deferred his doubts for the following day.

But the next morning was not really a happy one for Gattu. His mother looked uncommonly frail and crappy, prompting him to ask, "Are you sure you are ok? You don't look too good to me. I can stay back for the day, if you insist."

"Don't worry about me. I'll take care of myself. It's just the stress," she replied as she handed over a cup of tea to him, and carried on her routine with a studied calmness.

"If you say so," he said looking at her up and down. Her physical appearance gainsaid her assertion, but he couldn't force her into a confession if she was in no mood to confide in her. She had always been that way; fiercely protective and possessive about her privacy, sharing her discomfiture and pain only when it reached the acme of unbearableness.

He sipped his hot tea and deliberated. Should he ask her? Should he not? Should he ask her? Or not? Would his questioning worsen her physical well being further? Would it not? He dilly-dallied and then decided to ask, "What's this thing about me disappearing? Everybody seems to be talking about it. Was I really lost?"

She shook her head disdainfully. His short term memory lapse left her confounded as did the manner of his blinking 3 times in succession consecutively every time he looked at her. "We'll discuss it later. Not now. I'm tired. Too tired. I need to rest." she replied silently resolving to take him to the local doctor at the quickest opportunity to get his eyes tested. Likewise, Gattu too made up his mind to take his mother to the doctor. She appeared way too sick for his liking. He had an obligation to restore her health.

The local doctor declared her fit stating firmly that there was no reason for apprehension. All she needed was some rest to convalesce from the shocking kerfuffle of the past few days. The assurance did nothing to assuage Gattu's unease. His anxiety reached a crescendo when her fatigue worsened. The subsequent night, she was pronounced dead.

The trauma was unendurable. Nothing would ever be the same again. His brief life of 16 years was accustomed to the unknown, and yet the knowledge that the woman he loved the most, would no longer be a part of his existence left a queasy feeling in the pit of his stomach. He felt directionless, impuissant and stripped of all confidence.

But life waits for no man. Moments tick, memories fade, pains become less poignant and the traveller moves on. And so, did Gattu. He soon moved into the servant's quarters of the Master's abode. His jaunts to the jungle became more and more infrequent until they ceased totally. His 'blinking,' though, continued to hold a fascination for people but he himself suffered no discomfiture because of it.

***

CHAPTER IV

Separating a segment from the freshly peeled sweet sour tangerine, Dr. Memon looked at the nurse with a puzzled frown, "Are we doing all we can, I wonder?"

The nurse looked at him curiously. He sounded demoralized. "It's the best we can do, Doctor," she replied resolutely.

The doctor grunted, "I wonder," and put aside the remnant of the tangerine. "The cause is predicated upon the disease, but here, except for the unusual asthenia, there are no other symptoms. That's what has got me foozled. It's not just one patient remember, but several of them. I have been observing them aggressively based on the minutest of information that they have been imparting to me, but I admit to being defeated." He splayed out his arms in a gesture of ineptness.

The past few weeks had been traumatic and the events had started to affect him personally. He was tired of the liturgy in which the only sacraments were the medical reports which revealed nothing on which to base his prognosis. He got restive.

The nurse grimaced. Likwi, a sleepy village had suddenly awakened to the queer but tremendously shocking phenomenon of witnessing a spate of deaths for unknown reasons. Women died mysteriously after experiencing unwarranted fatigue. When almost a dozen of them met with a similar fate, the doctor got worried. Much to his dismay, he found his brilliance being threatened with the spectre of a failed diagnosis, and the unavoidable implications of the same.

After having spent a lifetime in a profession that was augmented out of passion, his intellectual self-esteem had taken a beating. He had been a bright student in college, and his days as an intern had been peppered by instances of brilliance constraining prominent hospitals to extend an invitation to join their faculty. Had he accepted them, he could arguably have had a flourishing practice in a more sophisticated environment, but he had declined; his sincerity and ardent love for the profession had induced him to practice in a village where his services could be easily accessed by the economically weaker sections of the society.

Reflecting uneasily at the unusual turn of events, he wondered if his catalogued breakdown of identification was misleading. He knew that a second opinion would be imperative if the disaster had to be averted - a view from an expert whose insight would cast another light on the situation. He dialled a number.

Dr. Smith was having his afternoon meal when he heard his personal phone ring. "I need your help," said a despondent voice at the other end. It was Dr. Memon, his friend and a man of integrity, who had always been revered for his opinion. He sounded irresolute. It was evident that the goings on in his place of practice had left him devastated.

The recent occurrences in Likwi had created ripples both in the media and the medical fraternity. The mental acquisitiveness was more so, because the dying patients were only females. Distinguished doctors outdid each other in putting forth hypotheses, each contradictory to the other. To be fair, some of them had even tried to finely tune their 'people sense' to advance plausible opinions, but the conceit of their medical qualification interfered in their ability to see beyond their advanced learning. As a result no concrete theories had been opened for debate, yet.

Like most of his brethren, Dr. Smith too had evinced a curious interest in the matter when it hit the 'News Headlines,' but his preoccupation had given him little time to ponder. Besides, Likwi was not so prominently placed on the map, as to warrant visits of distinguished doctors to test the veracity of it. He, like the rest, had, at that time, summarily dismissed them as rumours that would die a natural death over a period of time, but obviously that was not the case. Why else would Dr. Memon call him? After all, he was a doctor of distinction who excelled in a line of treatment that was never similar for 2 patients suffering from the same illness. His ability to read people and use that knowledge to heal his patient was so unique, that it set him apart from the rest of his peers. So, when he heard his voice at the other end, Dr. Smith knew that this was no longer a game. He took the first flight to the nearest city, and then motored down all the way in a hired taxi.

Dr. Smith was worthy of his calling. A man of few words, he guarded his profession as a prize that he had won, practising long hours and treating every patient with an undying zeal. He was exemplary, and unlike his other medical associates who drove past a scene of accident fearing legal problems, he always stopped by to offer his service. Every life that he saved was an award he cherished.

Both Dr. Memon and Dr. Smith had graduated from the same Medical Institute, but Dr. Smith had been a year senior. They had hit it off from the very beginning because of their shared vision of a healthier world and their passion for reading. They had parted ways under inevitable circumstances but continued to stay in touch on a regular basis.

When Dr. Smith entered his clinic, Dr. Memon was not in his seat. The place was nondescript and shorn of anything even remotely resembling comfort. The seat was an upholstered wing chair that looked like it was hardly sat upon. The walls were white and spartan, decorated only by a few credentials that declared the proficiency of the man who owned it.

Dr. Memon could have easily had his pick in any of the urbane commercial districts of the world, no matter how considerable, ostentatious or expensive, and yet he had chosen to stay in this insular locality. Dr. Smith's admiration for him increased twofold. As he kept running and rerunning his thoughts on the greatness of this humble person, he was interrupted in his deliberations by the familiar voice of his friend, "... and as with the previous cases, ensure that she is isolated from..."

The possessor of the voice stopped, his mouth opening wide at the pleasant surprise that awaited him. "You?" he almost yelled, as he spotted Dr. Smith. He sounded relieved, like a man in a desert who has found an oasis. It took him just 3 quick strides to reach his side.

Over a cup of hot piping coffee, they exchanged pleasantries during which Dr. Memon filled Dr. Smith in on all the critical details of the palavers that were making global news. He paused, and then concluded his narrative, "And now we have just admitted another patient suffering from similar symptoms. You may have a look at her, but later; only after you have rested. It has been a long tedious journey for you and I'm sure you are tired, but I'm grateful for your time. This place is a far cry from the modern settings that have become a part of your culture, but I have tried to give the best that this unostentatious town can offer. I apologise for any inconvenience that you may be put to, and sincerely hope you find your stay comfortable."

Dr. Smith was clearly embarrassed. "My comfort is the least of my concerns and you know that only too well. Let's not be agitated over a minor issue like that. Having a look at the patient is of overriding importance. Is she awake?"

"Yes."

"Let's go."

Examining the patient with exceptional gravitas and leafing through the reports that were kept in her room, he concluded, without the slightest hesitation, that the haemoglobin levels were healthy and the platelet count normal. He noted with satisfaction that the WBCs were doing their job perfectly well, the heart was pumping to perfection and the immunity levels were acceptable. The patient had no history of either diabetes or blood pressure; no thyroid problems and no anaemia! This was completely incomprehensible. Perhaps, there was more to it than met the eye. Like Dr. Memon, he too had a finely tuned sixth sense and an awareness of how to apply it, and like Dr. Memon he too unswervingly took the edge while diagnosing complicated cases where all else failed.

Turning to Dr. Memon, he said, "Strict chariness is the need of the hour. Monitoring her key habits and eating details are critical to the context. Experience has taught me that people reveal themselves in ways that, most often than not, go unnoticed unless they are aggressively observed. Ever so often, their habits contribute significantly to their infirmity. So, as I said earlier, the vigilance must be strict. We may have to conduct all the tests, once again, tomorrow."

Dr. Memon nodded in agreement. He looked at the nurse standing beside him and said softly, "Nurse, administer the drugs regularly and let the saline drip keep going... and yes, no visitors please until I give further instructions... and as Dr. Smith just advised... stringent attentiveness. Is that clear?"

"Yes," said the Nurse as she walked back into the Ward to administer a fresh bottle of saline to the patient. The doctors retired for the day with a lingering hope that the morrow would bring with it some better news.

***

Sharing a camaraderie that two men belonging to the same profession with a common objective enjoyed, the two doctors met again in Dr. Memon's cabin, the next morning, to review the situation. Morning salutations being taken care of, they had a quick breakfast consisting of hot coffee and some fresh local fruit. After discussing a few general things, they steered their conversation to the plight they found themselves in and to the whys and wherefores of it. They sincerely wished that the events unfolding through the day would be Panglossian, and more encouraging than the previous day's.

"Anything inappropriate or abnormal?" enquired Doctor Smith.

"Nothing, yet," replied Dr. Memon pressing his fingers onto his tired eyes that had awakened much before dawn when the night shadows still played with the corners of his room. He had prepared himself for the worst while hoping for the best.

"Good. I hope it stays that way. My last examination gave enough clarity to presume that she is on her road to recovery. Once she's well, we can get down to analysing and evaluating the source as well as the cause of this unusual dramatic turn of events."

But he had spoken too soon.

The nurse walked in to say that their most recent patient had said her good bye to the world. The stillness that dominated was discomfiting. Melancholia pervaded the air, and neither of the 2 doctors said a word, simply because they could not think of anything to say.

Dr. Memon's suppressed frustrations came flooding to the surface. He was not sure which feeling dominated - anger or self pity or an overlap of both. He attempted to break the suffocated soundlessness with a controlled voice but failed. "8 years of practice... 8 years of practice," he emphasized, "and nothing, to date, to undermine my integrity. But this..." He bit his lip and turned away.

Dr. Smith's silence expressed his understanding.

The past few hours had strengthened and deepened their bond to such an extent that Dr. Smith sensed the conflict in Dr. Memon's mind. He knew that saying nothing was crucial for this distraught moment, but words were necessary if he wanted to play a significant and hopefully dominant role in this delicate, sensitive situation. He stiffened himself and tried to look calm. His demeanour, he hoped, was propitiatory.

"I have never witnessed anything of this kind before; seems to be restricted to this locality. Since when has this been happening?" Dr. Smith finally asked, making a conscious effort to cover what he presumed was his ineptness which he thought was blazoned upon him for the world to see. He hoped his voice sounded realistic enough to conceal his lack of confidence and tentativeness.

"Of late," replied Dr. Memon softly. "It's something I have begun noticing lately. There's no antecedent to it; certainly not after I set up practice here."

"Is there a common thread that connects all of them?" Dr. Smith asked.

"Nothing that I can call to memory."

"Try remembering..." coaxed Dr. Smith

"I'm sure," said Dr. Memon. "Let's see... the weather... the air here is free from pollution... the food habits... no fast food joints, no cholesterol rich consumables... and the lifestyle... most of the natives have a commendable one. All of them are physically strong; none out here seems to shy away from hard work. There have never been any major health complaints... may be an occasional cold, cough and a minor bout of flu... nothing more. But now that you tell me—there is something common to all the patients who have recently passed away—their gender... all of them are females. But, that is common knowledge."

"Yes, that's common knowledge and the most reprehensible aspect." Dr. Smith paused. "Anything else?" he asked briefly.

"No nothing."

"Think hard... In the light of the situation, even the most insignificant thing may assume great significance."

"No, I'm sure," said Dr. Memon firmly. He hoped Dr. Smith had not perceived his uneasiness.

But Dr. Smith had. He had sensed his discomfort and concluded that being nervous was natural. He said nothing, because he wanted Dr. Memon to say something; just anything that would put a brake to the tension that was mounting, but there was no verbal response from him for a considerable length of time. Only his expression revealed the emotional blow he had been delivered. Finally when he did speak, it was to express a thought that was totally unrelated to the field of medicine.

"Do you think this necessitates something more than a medical intervention?" he asked, not because he believed in what he was saying, but merely because he thought some response was expected of him.

"Not sure," responded Dr. Smith. "I have no grave inclination or belief in black magic, if that is what you are referring to." He looked at Dr. Memon who nodded his head slightly to reaffirm that he was echoing his exact trail of thoughts.

Dr. Smith continued, "My only brush with the arcane world of sorcery had been when I was a child of 4 years, but I have been given to understand that it is rampant in this part of the world. I have irrevocable faith in our degrees, and yet we both know, through experience, that intellect is no substitute for pure common sense. This is one of those situations that demands innovation. I would like to stay in residence for another fortnight and feel around the fringes. Another case, and we could be on some trail or the other."

The conversation was abruptly stopped by an individual who ushered in a woman. He was a servant working at the Master's house. "She's complaining of the same symptoms that Leela experienced. Fatigue... tiredness..."

Dr. Smith swivelled towards him, half rising from his seat. He countered even before Dr. Memon could respond, "Nausea? Vomiting?"

"No nothing of that sort," said the servant.

"Since when has this been happening?"

"She complained of tiredness about 3 days ago."

"What part of the day? Morning or evening?"

"Afternoon."

"At what time?"

"Around 4 p.m. Maybe. I'm not sure."

"When did she go to bed the previous night?"

"I don't know. It must have been 11.00 p.m. That's the time she retires to bed, generally."

"At what time did she get up that morning?"

"The usual time, I guess; may be around 6.30 a.m.; the time when the kitchen gets busy. The Master likes to have his breakfast at 8 in the morning."

"Has that been her normal routine all these years?"

"For the past 3 weeks, as far as I know."

"And before that?"

"I wouldn't know. She's a new recruit... has been employed by the Master only recently."

"Who was there in her place before that?"

"Leela."

"Where is she now?"

"She died." The servant turned towards Dr. Memon and said, "Doctor Saab had attended to her case."

Dr. Smith looked at the doctor, "How did she die?"

"Similar symptoms," replied the doctor.

"Just like the recent patient?" Dr. Smith asked of Dr. Memon.

"I am afraid so," stated Dr. Memon. "Nothing was diagnosed. She experienced tiredness and fatigue for almost 5 days and then died mysteriously."

"How old was she?" asked Dr. Smith.

"Around 36," replied Dr. Memon blandly.

"And how long was she employed with the Master?" Dr. Smith continued with the questioning.

"For almost 15 years," answered Dr. Memon. "Very loyal, I must say. Besides being an exceptionally good cook, she also supervised the servants very efficiently. Was like a family member. The Master and his wife were heart-broken and devastated when she died."

"What about the 15 years that she was in employment with the Master? Any serious illness?" enquired Dr. Smith.

"No," asserted Dr. Memon. "She was very active, vigorous and full of beans. Her spontaneous enthusiasm had been very contagious, and her sense of humour was the talk of the town. Everybody I was acquainted with, loved her presence. Except for an occasional viral infection she was never laid up for any major sickness as far as I know."

Turning towards the servant Dr. Smith instructed him authoritatively, "Leave the patient here. She needs to be investigated thoroughly."

To Dr. Memon he said, "Let's examine her without any preconceived notions. All individuals are unique and so are their medical problems. They cannot be classified into specific categories. Conventional wisdom tells me that observations based merely on conscious signals would be misleading. It would be equally important to weigh the unconscious too, and convert that into usable perceptions."

* * *

CHAPTER V

Dr. Smith came from a family that was fairly rich, but rather conventional. His father was an independent financial consultant who was a great deal shrewder than he permitted himself to appear, while his mother was a devout Christian who was steadfast in her religious beliefs. His parents believed that a man was cultured only if he had, had the privilege of acquiring appropriate education. According to them, knowledge was a person's greatest asset. It was the only thing that could never be stolen. And to this end, they enrolled their children in the best of the educational institutions in the locality.

Dr. Smith had been a bright student in school. His thirst for knowledge accounted for his constant presence in libraries where he relaxed like one would at recreation spots in beaches. His voracious appetite for learning continued to be insatiable even today, as it was then. It had never dwindled. It helped to be well read and be abreast of the latest in the medical field. It was no wonder then that his patients unfailingly marvelled at his capacity for handling tricky valetudinarian cases.

And now as he looked at the new patient up and down, checked her eyes, her heartbeats, her pulse rate and the colour of her tongue as he had done with the previous patient, he was intrigued. 'There's nothing wrong,' he thought to himself. He couldn't quite remember when he had felt so unsettled the last time. Nevertheless, he had resolved to play a noteworthy, exceptionally professional role to enable a fruitful conclusion to the tragedy when he had received a call from his friend, and there was no looking back now.

He got down to the routine of a physical examination by asking her the obvious. Her exhaustion dismayed him. He would have to get a CBC done.

"Your name?" he asked.

"Acchamma..."

"Diabetes?"

"No."

"Let me check your 'pressure.'" She extended her arm. The systolic/diastolic in mmHg was 120/80.

"Normal. No reason for anxiety there. When exactly did you start feeling tired?"

"On Wednesday afternoon; around 4 p.m."

"Can you recount everything that happened to you until then?" He felt a little foolish asking her to recapitulate her morning, but he hoped to get some relevant clue that could lead him to something consequential.

She kept silent. Her puzzled expression suggested that she had not understood the question.

He decided to help her along with a few seemingly absurd questions. "O.K. let's start from the beginning. At what time did you get up in the morning on that particular day?"

"At 6.30 a.m."

"How did you know it was 6.30 a.m.?" Another witless question, he thought to himself, but he wanted to check if she was mentally alert.

"The clock on the wall said so," she replied. "Besides, the Sun was about to rise. From the kitchen window, I saw the red hues of the Sun spread like a carpet across the sky, but not the Sun itself."

"Wow!" he mumbled to himself, "had she been literate she would have been an excellent creative writer." Turning back his attention to her, he prodded, "Then?"

"I had a quick bath and went to the kitchen to rustle up some breakfast for the Master and his wife."

"Ummmmmmmm..."

"It was then that I realised that the tea leaves were almost over. There was just enough stock for the next morning's breakfast. I made a mental note to inform Gattu about it, later in the day."

"Gattu? Who is he?" asked Dr. Smith

"The Master's most trusted servant," Acchamma replied. "He's only 16, but excellent at keeping accounts. He has been entrusted with the duty of maintaining an inventory of the stock in the kitchen and elsewhere."

"Do you always approach him when you need anything for the house?"

"No, I don't," said Achhamma in reply. "It is Laxman who generally gives him a list of things we need."

"Now, who is Laxman?"

"He is the kitchen supervisor."

"Then why did you not ask him?

"He was not there."

"Where was he?"

"He was on a short visit to attend to his ailing mother."

"Continue," urged Dr. Smith.

Acchamma found it rather silly to narrate her daily routine to the doctor. But if he insisted on knowing, then perhaps he had a reason to do so. After all he was so learned.

"I kept the water for boiling," she continued in a voice that was beginning to waver because of the abnormal lassitude, "and beat up some eggs to prepare omelettes. While doing so, I peeped out of the window and noticed Gattu getting his bicycle ready. This meant that he would be going out on some errand for the Master. Unsure of his timing of return, I called out to him and requested him to come to the kitchen."

"Do you always do that?" Another ridiculous question thought the Doctor, but he was looking for clues in the unexpected.

"No, I don't," replied Acchamma. "I hardly speak to him. There's no occasion to, because Laxman handles all the groceries. I had never even met him despite working in the same premises. This was my first meeting with him. I was unsure when Laxman would return, and that is why I took the step."

"Oh! I see." Not that he could, but he was quite impressed at the way things operated in the Master's house. It was almost like the functioning of a corporate office. All the duties appeared to be well delegated. He made a mental note to visit the Master some time, if only to learn a thing or two from him. He brought himself back to the present as Acchamma continued with her narration.

"Gattu came to the kitchen and I told him that I would require tea leaves since there were hardly any left. He looked at me in disbelief. He wondered how the ration had run dry, a day or two earlier than usual. I told him that since there had been too many guests in the house during the month, the stock had got depleted. He seemed unconvinced but agreed, albeit a little reluctantly, to arrange for it as soon as possible."

"Interesting. Go on," encouraged Dr. Smith.

"I thanked him and reverted to my daily chores. And yes! I now remember something which I had relegated to the back of my mind. He blinked 3 times in succession quickly, as he looked at me. Simultaneously I felt a strange stab on my lower back. I thought it was a coincidence, so I looked at him again just to make sure that it was not; after all I was seeing him at such close quarters for the first time. At other times, we only greeted each other from a distance and that too only occasionally as my interaction was restricted to Laxman. Thereafter I met him several times, always in connection with some grocery requirement and at every meeting I couldn't help noticing the way he blinked 3 times in succession, in a weird manner, as he set his eyes on me. And yes! I did feel that inscrutable agonizing pain on my lower back every time."

"Oh, I see, and did you ask him why he blinked so."

"No, I didn't. I thought it would be impolite of me to do so."

"Did you discuss it with anybody else?"

"I did. I brought up the topic with Ramu, the other servant, who helps me in the kitchen. He said that it was a very recent thing. He used to blink quite normally until he got lost and was then found again rather miraculously after a few days."

"Anything else?" queried Dr. Smith.

"Yes, he also mentioned that except for the blinking of the eyes, everything else about him was normal. When I asked him if he felt any pain when he looked at him, he replied in the negative. That's when I concluded that there was probably no connection between the pain and his eyes."

"Are you sure?" stressed Dr. Smith.

"Yes. I'm sure."

"Besides Ramu, did you speak to anybody else about this aspect?"

"I did," said Acchamma, "but nobody seemed to find anything strange with my observation. They are all so used to it by now, that it does not intrigue them anymore. And about the pain; nobody mentioned it to me, so I guess it was sheer happenstance."

"Getting back to you, can you approximately figure out when you started feeling unwell?"

"Around 3.30 p.m. that afternoon, almost an hour after Gattu left."

"What exactly did you feel?"

"I suddenly felt weak; almost as if all the energy was being drained out of my body. I also felt an incomprehensible pain at my lower back."

"Were you overburdened with work, on that day?"

"No, I was not. The other servants were there to help me."

"Did you have your lunch on time?"

"Yes, I did."

"Keep talking."

"I told Ramu that I needed to rest awhile since my legs were feeling wobbly. I then retired to the servants' quarters. When I woke up I realised that it was about 6 p.m. Feeling a little refreshed I rushed into the kitchen; but then the weakness set in again. I retreated to the quarters. I have continued to feel weaker and weaker ever since."

Her voice had now dwindled to a whisper and she kept gasping between words. Dr. Smith realised that she was feeling extremely weak. "I think you should rest now," he advised. "I'll give you some sedative, so that you can sleep undisturbed."

He called out to the Nurse and instructed her to do the needful.

All throughout the examination Dr. Memon was silent. He kept standing just behind, and to one side of Dr. Smith. He had allowed the visiting doctor to take the lead and was happy that he had done so. Dr. Smith had examined the patient with an ease that was becoming of a doctor of his stature.

"Who is this Gattu?" asked Dr. Smith looking directly into Dr. Memon's eyes. "And what is this thing about him being lost and found again with blinking eyes?"

"Just a boy of 16 who works for the Master."

"Sixteen?" said Dr. Smith in total disbelief.

"Yes 16," said Dr. Memon. "I've always wondered about his age. His maturity belies that number, but his mother, while she was alive, was insistent on that. No birth records were ever kept before I set foot here. Now, of course, I've made it a compulsory norm. Of late, he has been a topic of conversation in this sleepy village. His mother too died of similar symptoms. Since the death of his mother he has been staying at the Master's house."

"And he is employed? Isn't it illegal?" asked a shocked Dr. Smith.

"Not here," said Dr. Memon rather casually. "Hard work is greatly appreciated, but funnily enough they do not take too kindly to school. Isn't that contrary? The local school teacher, though, is upbeat and buoyant of succeeding in his mission of what he terms as '100% LITERACY.' Let's wish him luck."

"Amusing!" laughed Dr. Smith. "And this thing, about him blinking 3 times in succession, every time he looks at someone... Sounds out of the ordinary, don't you think?"

"Yes!" agreed Dr. Memon. "He was never that way before. Apparently, he got lost when his mother was alive. At that time, he lived with her across the jungles, but now of course he stays at the Master's house. The landlord's men scoured the entire forest as also the surrounding areas but he was nowhere to be found. One fine evening, a few days later, he was miraculously discovered in the forest itself, unhurt, uninjured but with an enigmatic look in his eyes. Ever since, his blinking pattern has changed. Initially it baffled us... not anymore."

"I would like to see him," expressed Dr. Smith.

"Just an ordinary boy," said Dr. Memon indifferently, "but we can summon him to the clinic, if you so wish. Discounting his loyalty to his Master coupled with his 'blinking' there's nothing exceptional about him. But yes, he is very smart... very street smart and dependable... very, very dependable... so says his Master, the landlord."

* * *

When Gattu presented himself at the clinic that afternoon, Dr. Smith was engrossed in the reports of yet another patient who had been admitted with an undiagnosed ailment. He was visibly irritated when he heard a knock on the door and lifted his head to say

'Don't disturb me now. I'm busy,' but was stopped in his tracks by what he saw. The 'blinking' - it was grotesque, freaky and out of place on the young cherubic face that looked at him. Even from a distance it failed not to attract attention. Hearsay had psyched himself for the uncanny, but this was even beyond the periphery of his imagination. He wondered at the indifference of the locals. Had he been one of them, his fascination would never have ceased to be until he had got to the bottom of it. So immersed was he in his cogitations that he failed to realise that he was rudely staring at the entrant with his mouth agape.

"May I come in?" asked Gattu, once again, but this time a little more loudly.

"Oh yes! Come in, come in," said the doctor, composing himself in an ungainly manner as he ushered the lad into the cabin.

"So, you are Gattu?" He remarked, as nonchalantly as possible.

"Yes Doctor," replied Gattu, in all humility, standing all the while until he was requested to take a seat.

"How old are you?"

"Sixteen." The doctor looked at him in disbelief. He neither looked nor sounded 16, but the doctor refrained from altercating. He kept his counsel for a later time when he could discuss the same with Dr. Memon.

"I've heard that you are one of the brightest boys around the town. Is that true?" Gattu merely smiled demurely.

"I called you in connection with the patient who is presently under observation in this hospital."

The statement was far from the truth, and Dr. Smith knew that. He had been called in with the sole aim of satisfying Dr. Smith's curiosity, but obviously he could not tell him that.

"Since how long have you known her?" he asked of Gattu continuing with his pretence, while all along observing his abnormal 'blinking.' He kept searching for an appropriate word to describe it, but it evaded him.

"Ever since she took up duty at the Master's house," he heard him say.

"And before that?"

"No, never."

"It's a small place. Everybody seems to know everybody and everything that's happening around the town."

"True. But I've always lived away - beyond the jungles. I've only recently moved into the servant's quarters, after my mother's death."

"I'm sorry." The doctor bowed slightly in a show of condolence.

Gattu shifted restlessly in his chair.

"Was she always a sickly person?"

"Who?"

"Acchamma."

"I don't think so. I've never interacted with her except during the time Laxman had gone to visit his ailing mother in the neighbouring town. At that time, she approached me for a few grocery items and I complied. But I don't recall her complaining about her health."

"So, I guess you would not be able to impart any concrete info on her."

"I'm afraid not, Doctor."

"By the way, are you the same person who was apparently lost and found within a span of a few days?"

"Yes Doctor. That's what everybody says."

"Everybody? And you?"

"I don't remember anything."

"Isn't that strange? Getting lost and then being found is no small thing, and yet you say you are oblivious of such a happening?"

"It does sound implausible, but it's the truth."

"Memory lapse, you think? That's quite a possibility. Have you ever suffered from such mental misadventures in the past?"

"Never. My Master will vouch for that."

"That's odd! What was the last thing you remembered before you were proclaimed to be lost? Or has that moment also been mired in a maze of amnesia?"

"No," replied Gattu. "I remember it with distinct clarity... almost as if it happened yesterday. It was late... my thoughts dwelt on numerous issues including my mother and the warm savoury meal that would be awaiting me, when I became aware of a distinct red light in the horizon. It's a colour that continues to be vividly etched in my memory, as would a rare souvenir in a collection of a tourist's keepsakes. I thought it was a delusion, but common sense prevailed, and I looked at it again. It was real and appeared to loom from the densest part of the jungle. I found myself walking towards it. There was a weird buzzing sound which got louder and louder as I inched closer and closer. The next, I found myself at the Master's house with everybody making a big brouhaha around me."

"Fascinating. Did you start blinking this way, only after that?"

"Presumably so."

Their conversation was interrupted by the Nurse who entered to say that the incumbent patient was not responding to the treatment, as reckoned. Dr. Smith diverted his attention, for a moment, from Gattu and looked at her, "The saline drip? Is it still on?"

"Yes," she replied.

"Not to worry. Be patient. She'll get better soon," he assured her. Then as an afterthought and out of sheer courtesy he asked her, "By the way, have you met Gattu?"

"No," the nurse replied, looking peculiarly at Gattu who appeared ill at ease with, what he presumed was, unwarranted attention. "But I've heard of him." She turned to acknowledge the introduction with a simple 'hello' but instead found herself blurting involuntarily, astounded, "Why is he blinking like that?"

"Unusual, isn't it?" the doctor said. Gattu dropped his eyelids and looked at the floor beneath his feet. He wanted to get away from the scene as quickly as possible. Such reactions had become the norm in his unassumingly simple life, and yet it failed not to disquiet him. "May I leave now?" he asked of the doctor as a means of escape from the awkward scene.

"All right, Gattu, you may go now. We'll meet again."

"Thank you, doctor," said Gattu, exiting as quickly as he could.

"Delphic!" remarked the nurse, once he left the cabin. "And gut wrenching!" she added with her hands clasped on her lower back. The doctor gave her a sharp look. "Gut wrenching!" he proclaimed in astonishment.

"Yes! Not sure if it was coincidence, but I did feel a poignant spasm in my lower back when I looked into his eyes."

The doctor made a mental note of it. Dr. Memon who had gone out for a while walked in just as the nurse walked out. He sat himself comfortably in the chair opposite, but not before greeting the nurse perfunctorily.

"How was your meeting with Gattu?" he asked Dr. Smith.

"Upbeat! Have you ever met him?"

"Yes, quite often," Dr. Memon replied.

"What is your opinion about his blinking?" asked Dr. Smith. "I thought it was very unnatural. Paranormal is the word. Just like a keyed doll that starts blinking the moment we turn the key. Intriguing! I think I ought to meet him again. And another thing! The nurse says that she felt some pain in her lower back when she looked at him. Remember she has met him for the first time and she's a female. That's some food for thought."

Dr. Memon brushed aside the doctor's last comment with a careless shrug and was about to give a reply when the nurse came running in,

"Doctor, doctor," she said hysterically, "the patient has died."

"What?" He was flabbergasted; there had been no alarming indications that could have led to such a situation.

* * *

It was sensational, a phenomenon that had no precedent for its ill-omened notoriety. The media was on an overdrive. 'Likwi,' an unknown, insignificant place had unexpectedly hightailed to occupy maximum bytes for its opprobrium. The experts, the opinion givers, the ordinary citizens, all had their independent suppositions with no hard facts to support their hypotheses.

The newscaster read the news with a dead pan expression and in a very pedestrian fashion,

"... The village of Likwi has been witness to a series of deaths for which medical science has failed to assign a legitimate reason. All the dead coincidentally are females. The local governing council has expressed grave apprehensions on the issue. The decision has manifested itself in the appointment of a high level investigative team that incontrovertibly identifies with the highest professional and medical values. They have declared an emergency and intend to take it up on primacy over other matters. Hopefully it will reach a viable conclusion and assist in reversing the spiral of rising deaths which show no signs of abating...

... in a bid to solve the impending disaster, not of its making, the local council is now embarking on a hardnosed approach..." the monotonous voice carried on.

Rishi switched the channel to the Sports Section and looked at Alex questioningly, "What do you make of this?"

"Referring to Likwi?" countered Alex. "Life is certainly not ecstatic there, definitely not an enviable place for women who seem to be vanishing, most systematically, one by one."

"Yes! Ostensibly, there's no justification to support the innominate deaths," stated Rishi.

"What do the reports say?" asked Alex.

"Nothing relevant that the doctors can place their fingers on," replied Rishi. "The victims were all evidently fine and kicking, with no past medical history until they kicked the bucket following a host of inexplicable prodromes. The ailing ones continue to breathe while the indecipherable infirmity takes a toll on the healthier ones. The fact that all of them are females makes it even more horrendous."

He tried, distractedly, to concentrate on the morning newspapers freshly delivered to him. The headlines were similar–'THE LIKWI ODDITY,' as nicknamed by the Press, dominated the front pages of all the prominent newspapers. At a time when 'women empowerment' was gaining momentum here was a place that mocked the world by behaving to the contrary. The situation was worrisome - vacillating between the inconclusive and the monstrous.

His thoughts were intruded upon by Alex. "Is that a male-dominated society?"

"That is the ironical contradiction," said Rishi. "It's one of the few localities in the vicinity that has been encouraging the birth of a female child. And even if we do consider for a moment, that it is a male-dominated society, why would adults die? The target would have been new-born females or female foetuses."

"Gracious me!" exclaimed Alex. "There's something menacing underlacing the eventualities. I would be keen on knowing what is happening. What is the latest, by the way?"

"The local nurse has just died of similar symptoms."

***

Dr. Memon wasn't the only one. Dr. Smith, too, was visibly perturbed. He was uneasily conscious of his poor attempts to pacify Dr. Memon which was the least he could do at this juncture. "I understand your feelings of guilt," he said, "but let me assure you, they are misplaced in the context. I have known her only for a brief period but your association with her has been for a fairly long time. Are you sure she never complained of exhaustion before?"

"No, never," replied Dr. Memon. "Not in my wildest dreams did I ever imagine that she would be a victim."

"Let's recall the flow of events," urged Dr. Smith. "She was fine when Acchamma was brought in for treatment."

"Yes, she was perfectly fine then," agreed Dr. Memon.

"Did anything untoward happen to her before Acchamma visited us?" prodded Dr. Smith.

"No," stated Dr. Memon. "She had attended to every single patient who had been admitted here."

"Meeting with Gattu? Has that happened before?" posed Dr. Smith.

"No," said Dr. Memon. "This was the first time that he was called to the hospital. She had once remarked she was curious to meet him. That, however, never happened until you came. Thereafter she got to meet him not once but several times but always in our presence. She did tell me much later that she failed to understand the backache she experienced every time she looked at him. I dismissed the statement summarily, since we had also met him at the same time and had not experienced anything similar."

Dr. Smith reacted with a quick jerk. "Tell me one thing," he said, "Did all the women who died suddenly, meet Gattu at some time or the other? The nurse did, and I know Acchamma did too. The others? I know not."

"I know not, either,"

"Then we should find out."

"I don't see the connection, but we'll find out if you are interested."

His voice took on an affirming intonation when he said, "Just something that struck me. Of course! It's my view at the moment."

* * *

Everything was panning out immaculately–exactly as they had envisaged. The Martian territory basked in their success.

"The whole kit and caboodle have stitched to consummate perfection. Just as we had anticipated. We must now move swiftly on to the next part of the Plan - focusing on other parts of the planet to increase the number of anonymous, unmarked female corpses expeditiously. That's the paramountcy," said Pontus smugly.

"Then why the dawdling?" asked Maxus.

"'Twas appurtenant to assess the ramification. This unexampled triumph will now fifth-gear it. In the locator next is the city of Hosuru, supposedly the most populous place on Earth. People, here, live in tiny units purportedly jostling with one another for easefulness."

CHAPTER VI

The Sun was setting. Mrs. Sheen looked out of the window and saw the last red of the sky sink behind the tall building which had come up recently. The city was growing in density and getting too crowded for comfort, she grumbled softly. Ten years ago, when they had set up home here, it had been a delightful place. Small bungalows had lined the boulevards, and the down to earth Pop & Mom stores had been offset with tiny parks.

Now the gestalt of the place had become inaesthetically dotted with ruthless concrete buildings that touched the sky. Huge cold shopping malls and traffic choked avenues only added to its unsightliness. To say that it had altered beyond recognition was certainly not an exaggeration. Even the redness of the sky seemed to have changed, especially so today. It was a mix of scarlet and crimson submerging into a grey fogginess. Strange! She had never noticed it before, or possibly she had never really paid so close an attention to it, in the past.

Picking herself from the chair, she braced herself to prepare the evening meal. Being a homemaker was not her idea of an interesting life. She would have much preferred to pursue a career. But she had conscientiously given it up after the birth of her son, and deferred it to a time when he would no longer need her presence. Until then she would have to do things that were not necessarily to her liking.

It was almost 7.30 p.m. when she entered the kitchen. She was pondering on what to rustle up for supper when the door bell rang. "Must be Anil," she said softly to herself as she opened the door. It was not - it was his father.

"Anil? Not home, yet?" he asked.

"No. I expected to see him at the door, and not you. He should have been home by now," she said looking up and down at him. He appeared tired. "But you're early. How was your day?" she asked.

"Good, but a little tiring," replied Mr. Sheen nodding his head absent-mindedly and preparing himself for a quick shower. His son was intellectually sharp and academically bright. He didn't need additional coaching, but had insisted on it, just to be with his best buddy, Abhishek. As a twosome, both Abhishek and he partook of a lot of activities together... they attended the same school... enrolled for the same swimming sessions... frequented the same guitar classes... and now even went to the same coaching class together.

He wondered if he was pampering his son needlessly. His own parents, had they been alive, would have advised him against doing so, saying that such mollycoddling would spoil him. They had been very strict with his upbringing, and he had to admit that it had done him a whole lot of good when he grew up. Discipline was necessary not only for a successful career, but also for a satisfactory personal life. But Anil was his only child and saying no to him had never been easy. Besides, he had proved to be a brilliant student, so there was really no need for any regrets.

The aroma of food drifted through the hall and titillated his nostrils. It was dinner time - a time when the family got together to discuss the trivialities of the day. He hummed softly under his breath and throwing a flannel towel across his shoulders, entered the spacious bathroom. Anil would return by the time he had finished his bath. But when he walked out refreshed from a warm-water shower, his son had yet not come home. His initial smugness turned a bit anxious at the edges.

"Don't you think it's a little too late," he said rather apprehensively to his wife. "I think we should call up the teacher."

His wife agreed. She had presumed that the teacher may have detained him for a thorough revision. He had his history test the next day, a subject he disliked because it involved a lot of cramming of facts. But that was no reason for him to be so late. She wondered at his unusual lateness. The tutor was a responsible person who knew how paranoid most parents were, and would not retain the students beyond a particular time. She was sure he had not gone to Abhishek's house either. He would have certainly informed her, if he had done so. He always did. It would be best to check and put her anxieties to rest. She dialled the tutor's number. The voice at the other end responded with a cheery hello.

"Good evening Ma'am. Have the students left from your premises?"

"Who's on the line?" the teacher asked.

"I'm Anil's mother."

"Oh hello! Yes, yes! The students have already left from here. They left quite a while ago. Why? What's the matter?"

"Anil's not reached home yet."

"Oh! Then why don't you check with Abhishek? The two of them left together."

"Yeah! I'll do that. Thank you, so much. I apologise for disturbing you at this hour."

Mrs. Sheen turned pale and broke into a sweat. Her lips were dry, her expression forlorn.

"What's the matter?" enquired her husband.

"He's not at the teacher's place," she stammered. "We'll have to check with Abhishek. She says that the two of them left together."

When the landline buzzed, Abhishek's mother was laying the table for dinner. She was mildly astonished when Mr. Sheen uncharacteristically asked for her son, without the preliminary salutations. He was always so polite that his brusque attitude took her by surprise. Perhaps something was wrong.

"Abhishek!" She called out to her son, rather impatiently.

"Yes, mama!"

"Uncle Sheen is on the line. Please talk to him. He sounds disturbed."

Abhishek came running out of his bedroom and pulled the receiver from his mother's hand, "Good evening, Uncle!" he said, and replied in the negative when asked if Anil was with him.

"Then where could he have gone?" said Mr. Sheen with a frenzied plea in his voice. "The teacher said that the two of you left together. Any idea where he could be?"

"With Uncle Dhruv, I guess. We met him on our way back. He offered to give me a lift, but I declined because I had almost reached home."

"Is that so?" There was soulagement in his voice. Turning to his wife, he said, "I believe Dhruv offered him a ride on his bike."

Mrs. Sheen relaxed visibly, but was furious with her brother. He was a bachelor, and was irksomely irresponsible. His erratic habit of visiting people unannounced, had always been a bone of contention between them. As usual, he had probably decided to spend the night with the family, and while on his way to their house given a joyride to Anil. It was time she had a word with him on social behaviour and etiquette. Not that he'd lend her a patient ear, but she would persevere. Or maybe she would just get him married. Marriage had a sobering influence on individuals and made them more accountable. But where was he now? It was late, already too late.

She dialled his number.

"Here, let me try," intervened her husband, when she failed to get him on the line for quite some time. A good 15 minutes later, the phone rang, and Dhruv came on the line, but only to express shock and concern. "How disconcerting! I'm sure there's some mistake. I'm in Naleru right now, and not at Hosuru. Besides, I don't possess a bike anymore. I sold it, and bought a new car for myself. As usual, I forgot to tell. How stupid of me! I'm sorry!"

"This is unbelievable!" Mr. Sheen exclaimed. "Abhishek was quite certain it was you. He has interacted with you so many times that I'm sure he could not possibly have been mistaken."

"I'm not trying to contradict him," replied Dhruv. "If he did say that it was me, it may have some truth in it. But how could it have been me? I'm here."

"Who do you think it was then?" asked Mr. Sheen.

"Wish I knew," declared Dhruv. "There's some misunderstanding. Get the matter clarified immediately. And yes, one more thing, call me if you need anything from my end."

"Sure," replied Mr. Sheen in a whisper, as the wrinkles of worry deepened on his forehead. His fingers trembled as he looked at Mrs. Sheen disbelievingly, "Dhruv is in Naleru."

He sounded more like an uncharged robot that had to drag itself through a vacuum because it lacked both a battery and an updated menu to spur it on.

"What?" Mrs. Sheen screamed hysterically. "Call Abhishek. Maybe it's a case of mistaken identity. Here, let me talk to him this time."

Her tone failed to conceal her fretfulness as she coaxed Abhishek into remembering, "Are you sure you saw Dhruv on the motorbike?"

"Yes Aunty," replied Abhishek confidently.

"But Dhruv is in Naleru..."

Abhishek insisted, "I am certain it was Uncle Dhruv. I even spoke to him." He was unflappable in his admission.

The telephone conversation ended irresolutely, and was followed by a seemingly endless deceptive calm during which Mrs. Sheen attempted to weigh a fruitless problem for which she could find no answer. Nerves itching, she tried unsuccessfully to dismiss her diagnosed fears and glanced at her husband, "Something's not right. We need help, and the police would be better poised to offer us that."

Mr. Sheen muttered inaudibly in agreement.

"Carry a photograph of Anil," he reminded his wife as he buttoned his shirt. "It would be required for identification. And yes, a photograph of Dhruv, too."

"Dhruv?" A flicker of astonishment rose and died in his wife's eyes.

"According to Abhishek, Anil was with him... the last piece of information that we have of Anil. His photograph would assume significance in that context."

Thoughts flowed incoherently, and their state of emotions clogged into a still, solid opaque ball steering them towards only one direction... the police station. The unabated ringing of the telephone, and the soft music on the radio symphony echoing in the background added to the gravity of the situation. A response to the shrill of the phone was unwarranted and could be postponed to some other time; as for the radio–it could be switched off, later. The tragic situation of their 'missing' son had relegated everything else to nothingness.

When the Sheens entered the compound of the police station, the Inspector was sitting astride his two-wheeler. After a gruelling day at work, he was getting ready to leave for his home. He glanced at the couple with a slight lift of his eyebrows, his eyes moving from one to the other in rapid succession to grasp their general aimlessness as they moved towards him. Obviously, something was wrong.

"What's the matter?" he asked in a tone that was severe but solicitous.

"My son, Anil," stammered Mr. Sheen. "He's gone missing."

"Gone missing?" countered the Inspector.

"Yes. He has not returned home," replied Mr. Sheen.

"Ah!" The Inspector retorted with a snigger. It annoyed Mrs. Sheen to hear the Inspector's contemptuous chuckle. She closed her eyes and compressed her lips in controlled anger, but failed not to pay close attention to his routine questioning.

"How long has it been, since he supposedly went missing?" she heard him ask her husband.

"Almost 2 hours, now."

"Just 2 hours? And you are worried?" There was a hint of derision in his eyes. He noticed the shocked swiftness with which the couple looked at him. "I am not sure that there is any cause for worry," he continued with a disinterest that was unbecoming of a man of his authority. "It is so like his generation to be loitering here and there. He'll be back any moment."

"How can you be so callous?" screamed Mrs. Sheen making no attempt to conceal her bitterness. It was difficult to grasp that a man of his calibre could display such caustic rancour in contexts that needed sympathy, and not acute indifference. The Inspector looked at her. He was aware that in situations like this, realism was usually equated with affected speech, but he belonged to the no-nonsense category that preferred to call a spade - a spade.

"Ma'am, I sympathize with you," he said in a propitiatory tone that he generally reserved for women. "I know it's not easy for a parent to lodge a complaint for a missing child, but such stories have ceased to astonish me... I have become impervious to them. We solve such cases every day, and they are all similar... a kid is lost for a few hours...only to be found soon after. Most of them are single kids who are restless and bored. Staying at home, all alone, while their parents are at work is not quite a palatable choice for them."

"With all due respect to you, Sir," intervened Mr. Sheen, "I would request you to refrain from uttering such derogatory words, and concentrate on the solemnity of the occasion. And yes, let me tell you one more thing - my wife does not work. She stopped doing so after our child was born. The decision, to give up her career temporarily, was mutual."

The Inspector shrugged; a shrug that was an emotional equivalent of a statement.

Mr. Sheen continued in a voice that reeked of annoyance and sat down in a chair without waiting for an invitation to do so. "You have to understand Inspector that we have come to seek your help, at this hour, because the problem is beyond our limited understanding. Abhishek, our son's friend, told us that he was picked up by his Uncle Dhruv on his way back home."

"Then what's the problem?" countered the Inspector.

"We spoke to Dhruv. He says it was not him," replied Mr. Sheen.

"Did he give an explanation for his denial?"

"Yes. He's in Naleru, and not in Hosuru. He went there last evening on an official visit."

"Does Dhruv have a twin, by any chance?"

"No, he doesn't, and before you ask me if he has any other siblings, let me tell you that he has only one, and that is my wife," stated Mr. Sheen firmly.

The Inspector straightened his back. "Who is this friend? And what did you say his name was?"

"Abhishek."

"Well get him on the line for me. I would like to have a word with him."

After a brief conversation with Abhishek, the Inspector stared into nothing in sheer disbelief. "Have you got a photograph of Anil with you right now?"

"Yes," said Mr. Sheen. Removing the photograph from his pocket, he showed it to the Inspector who amidst interposed remarks like "Goodness gracious!" and "How sad!" slipped it into his shirt pocket. "And what about this Dhruv? Do you have his too?"

"Yes."

"Good." The Inspector's expression sobered. "We'll get cracking on the case immediately. In the meanwhile we can start with filing an FIR for a missing complaint. Any idea or suggestion to offer? For instance, do you have enemies? Do you think it may be a case of kidnapping for ransom? Please do not conceal anything from us. We need your total cooperation to get to the bottom of the case. And yes! I'm sorry for what I said earlier but you must condone our attitude. We have visited and revisited such scenes so many times in the past that we have become quite cynical. I thought it would be all so simple and logical, and therefore so manageable. Apparently, that's not the case here. But I agree, that's no reason to be insensitive."

The Inspector's apologies were genuine.

* * *

The euphoria that reigned in the Martian territory was subdued. Orchestrating the pickup of a specimen from amidst a moving populace was not easy. The place was crowded and condemned to constant watchfulness, and the job had to be clandestinely done to avoid drawing attention of the passers by. 'Twas a 'cloak and dagger' affair. The last one had been comparatively easier because of the inscrutability and sparseness of population... almost like child's play... a fortuitous operation largely an accident of geography. Not here, where there were no desolate moments. To avoid far reaching consequences that would shatter the myth that the Earthlings had about their planet, the seizure had to be immaculately executed to mask it from the critical appraisal of the inhabitants.

"Picked up the new specimen?" asked Maxus of Pontus as he stepped into familiar terrain.

"Yes."

"Any hitches?"

"None that we had prepared ourselves for," proclaimed Pontus. "The task was surgically precise, almost fine tuned to perfection, but, as anticipated, arduous and incredulously complex... too many pairs of curious eyes all around; but the vision of 'Earth-control' steered us through. The Earthlings may not be blessed with a mental prowess of our magnitude, but they do have an alert brain that sometimes, most bewitchingly, manages to fathom a lot of things. They call it the sixth sense, I think. Besides, of late, they are in possession of a crude invention called the 'Television' which is a conduit for information. One mistake, and the whole of Earth would know of it. 'Caution' was the password to avoid detection. But one thing's for sure, this exercise has confirmed what we had experienced earlier."

"And what is it, if I may ask?" posed Maxus.

"Self-indulgence has utterly taken over and their complacency has become very predictable. It's a world where perceptions count rather than facts," opined Pontus.

"Contradicting yourself again?" derided Maxus. "In our earlier conversation you did argue that they have become dangerously rational and digital."

"I still maintain that," asserted Pontus, "but certainly not to the extent that I had presumed, but that stood at an advantage for us and always will."

Maxus looked at him questioningly.

"If you recall, we had enacted our moves repeatedly to acquaint ourselves with every manoeuvre to avoid last minute glitches. That served us well. The ruse was executed to the letter. Of course, we had replacements in mind should anything untoward have happened. But nothing did."

"Then what prompts you to comment on their complacency?"

"Well! As we whizzed past the curve where the gravitational pull was at its peak there was a perceptible change in the atmosphere. Had it been us, we would have been spurred into action on noticing a sudden extraordinary redness, and an inexplicable hazy smog that lowered their visibility to almost zero - but not they. There were a few murmurs, a few indifferent dismissals of the shoulders and a sense of relief that they were all safe."

"And the human sample?"

"He was the easiest. Just a moment's hesitation before he accepted the ride. We had not erred in duplicating a robot to the likeness of his close relative. They were fooled and how!"

"They?" Maxus looked at Pontus interestedly.

"Yes," verbalized Pontus. "He was with a 'friend'– to use the lingo of the planet. The location was deliberate; to extricate a negative response from the supposed friend who lived a negligible distance away from the spot."

"Splendid!" acclaimed Maxus. "It's becoming increasingly clear now that it's not the Earthlings that we need fear as much as the Venusians. This particular exercise corroborates that. Pulling a fast one on these creatures of the Earth is easy, but those dreaded elements... they are a different cup of tea altogether. But then, both you and I have always maintained that."

"Oh no! Don't bring that up again," postulated Pontus. "I agree, there can be no room for dissent on their infuriating interference. But I wouldn't like a discussion on that, at least not now, to be a spoiler of our moment of triumph, transient as it may be."

"I wouldn't either," stated Maxus. "As much as I try to relegate them into the background, their existence keeps hovering in the outskirts like fruit flies around a bowl of ripened fruits. It would be inopportune to entice them into a web of scepticism. Anyway, let's disregard them for a while and focus on the new specimen. How is it?"

"Not as raw as the earlier one," said Pontus.

"What am I supposed to extrapolate from that?" queried Maxus.

"The earlier specimen had been to a local makeshift place of teaching," elucidated Pontus, "and that too occasionally. His brain had never been satisfactorily educated to fit into the existing arrangement of things on Earth. Often, it was attuned to follow instincts, but when commanded by those around him, it obeyed unquestioningly. This one's been to a regular school."

"School? Now what's that?" Maxus was disgusted. It was about time they prepared a document on the vocabulary of Earth. It would be useful. New words always cropped up in his conversations with Pontus and he appeared to be wasting a lot of time deciphering their meaning.

Pontus proceeded to explain, "It's a place where a supposedly knowledgeable person with an apparently great mind and great productive ability, is given the responsibility to mould young humans, referred to as 'students,' into clones. In an attempt to sharpen their brains, such 'mental duplicates' are taught in a solemn, severe tone to think, question, rationalize, memorize and analyse every experience they encounter. In the process most of them lose the creativity and originality that they were born with. In the stratum of society where he belongs, if an individual doesn't attend school, he or she is referred to as illiterate and dim witted."

"Is that a shortcoming or a virtue?" enquired Maxus.

"Both," said Pontus distinctively.

"I fail to understand," acknowledged Maxus.

"Since the brain has already been trained to consider knowledge as anything but superfluous, it considers judgement elementary before reacting," illuminated Pontus.

"That statement is a wee bit complex. Maybe you could make it simpler with a segmented explanation," pleaded Maxus.

"It isn't," explicated Pontus. "All the same let me rephrase it for you. The brain of the earlier specimen was not habituated to questioning. It obeyed orders when given, and acted impulsively when not. Not this one. It has been schooled to think in a specific manner. The thought process is more advanced. It pauses, questions and again pauses before it acts or reacts. But once convinced, it accepts and obeys unquestioningly."

"Does this imply that we will have to handle him a little differently?"

"Yes, the chip may have to be altered for a few codes."

"Have we worked on that?"

"We have."

* * *

He reached for the light when his telephone buzzed. It was 7.00 a.m. The bedside clock told him so. He glanced at his wife who seemed to have been undistracted by the sound of the phone. Her face wore a tired look - a sign that her enervated body had probably just fallen asleep out of sheer exhaustion. He did not wake her up. 'Peaceful sleep' had evaded both of them ever since their child disappeared mysteriously from their life. He smiled sardonically when he heard the Inspector at the other end, "We have found someone who looks like your son and we need your presence to identify him." The voice was crisp, very inconsistent with the gross manufactured persona which Mr. Sheen presumed was deliberate. He had been right in sensing that the Inspector's derisive disposition cloaked a smart and practical individual.

Quickly getting into his trousers, he slipped into his sandals and walked the distance to the Police Headquarters reflecting on the past few days that had left both his wife and him totally devastated. The situation had been remarkably different as compared to those in detective novels. There had been no frequent disruptions in the form of constant visits by investigators, and no tiresome gruelling sessions with the police who had been admirably sensible. They spoke to him only when necessary.

The media had made a mention of it, but the coverage was not as wide as envisaged. The episode had been deliberately downplayed for fear that too much publicity would deter the authorities from doing their job proficiently. Undue attention was uncalled for in a situation that sought solutions rather than hype.

"Yes, it's him," he said, 10 minutes later, in a voice that sounded shallow to him. He kept staring down at the familiar face of his son in disbelief.

"Where did you find him?" enquired Mr. Sheen of the Inspector.

"In the local garden, on a bench."

"What?" Mr. Sheen exclaimed incredulously.

"Yes, you heard right."

"But... but," stammered Mr. Sheen. "But, I thought you said you had searched all the plausible places, in and around the neighbourhood, and had been unable to trace him. As far as I remember, you had even discussed the possibility of him being kidnapped and taken to a destination outside the city."

"Yes, I did," confirmed the Inspector. "But that was an inconclusive colloquy, to be carried forward at our next meeting. We ourselves were nonplussed when we found him there. We least expected to."

The Inspector then went on to narrate the details as they unfolded, "I was having my morning cup of tea when my men called to say that a boy resembling Anil was lying in the garden. Donning my cap, I called for the official transport and accompanied by 2 other plainclothes policemen manoeuvred our vehicle towards the park. Walking impatiently up to the constable who was waiting for me, I asked him if he was sure it was Anil. Not that I doubt the credibility of my staff, but it's best to double check. I then gave the child a second look. I realised that he did resemble the picture given by you. That is why I called."

The Inspector paused significantly and gestured to the constable, on duty, to resume from where he let off.

The police officer carried on from where the Inspector had left off, "I was on my routine round of the city when I noticed a young body from outside the precincts of the parkland. My first reaction was to ignore. It is common for a few homeless ruffians to take shelter here in the dark of the night. But my sixth sense nudged me to explore further. Getting down from the vehicle, I walked towards him. I was astounded. That's when I decided to call Inspector Saab."

Continuing the narration, the Inspector said, "I was aghast, to say the least. My mouth remained open in astonishment for quite a while. I can say with certitude that there had been no trace of him when we searched for him here, the first time. It is paradoxical that the very first place we had scoured is the very spot where we found him. I remember taking out my big white handkerchief from the pocket of my trousers and wiping my face with ferocity. I hoped my action would hide my mental agitation, but it was not of much help. Anxious and confused, I looked at him again, and then again with cynicism. I even requested the constable to give him another closer look. He did. It was then that I dialled your number. Now that you have identified him, let's complete the formalities and bring a closure to the case."

The Inspector removed a sheaf of papers and placed them on his desk. "And yes, he needs to be attended to immediately, by a doctor. We have called for one."

Mr. Sheen looked at his son again. Except for the slight rising of the chest spasmodically, there seemed to be no visible movement in him. He tried waking him up, "Anil, Anil! Get up!"

"No use," prompted the Inspector. "He appears to be sedated. We did the same, but with no avail. As I said earlier, he needs prompt medical attendance. The doctor is already on his way. He should be here soon. His advice would be vital to understand his physical well being. In the meantime, I request you to complete the paper work of the constabulary."

The Inspector knew he sounded insensitive, but when a man in his profession became sentimental it would be time to quit.

Mr. Sheen felt a momentary sense of depression. He had to admit that the Inspector had been more humane to him than the context warranted. He couldn't help but admire the confidence with which he displayed a total command of the situation. Very few men in authority carried their responsibility like a towel thrown round their shoulders in the most casual fashion, and he was certainly one of them, but his bolstering presence was not enough.

He missed his wife sorely. He wished he had brought her along; but he had, had no heart to disturb her. She would have had her moments of hysterics; but her habit of coping would have soon taken over. For all her excitability and her frivolous habits of inane spending, she had a core of sane solidity which assisted him in staying happily married to her.

The Inspector continued to make cursory notes on a thin writing pad, more as a matter of formality than anything else, and ran his signature across relevant spots to indicate the closure of the case.

The doctor who walked in soon after, examined the patient with alternate utterings of, 'oh,' 'very strange,' 'unnatural' and concluded his examination with an air of finality. "Everything's normal," he stated. "There's no fever. The pulse is regular and so is the heartbeat. No sign of any external or internal injury either. He only appears to be asleep. He will wake up in a few hours from now," he said with a firmness that caused the apprehensions of those around him to subside temporarily.

As an afterthought he added, "I may, however, have to re-examine him, once he regains consciousness," he reflected ruefully, unsure if he was right in his understanding of the case.

"What exactly brought him into this state?' the doctor asked looking at no one in particular.

"He had been missing for a few days. He was nowhere to be found - not until this morning when we stumbled upon him in the local garden. He was lying there, unconscious, on the bench," said Mr. Sheen.

"Interesting! Sounds more like a cock and bull story to me," the doctor said disbelievingly.

"Agreed. Sounds implausible and enigmatic, but that is the sad truth," intervened the Inspector.

"You're right. It's unbelievable, but there's nothing to show that it may have been a forceful abduction. I see no abrasions, or for that matter any other sign of bodily harm. He appears to have been well looked after, wherever he may have been. Anyway, trivialities aside, I reiterate that I detect no abnormalities. You may take him home for now but be sure to present him at my clinic as advised earlier," said the doctor with total authority.

The preliminaries at the Police Headquarters being dispensed with, Mr. Sheen took his son home for a much-needed rest.

It was well past midnight when Anil finally opened his eyes, "Where's Uncle Dhruv?"

His mother who had not left his bedside for a single moment, called out to her husband excitedly, "Listen, Anil has regained consciousness, and he's asking for Dhruv."

Mr. Sheen entered the room with a look that gave the impression that he had been trapped in some intrigue, and had only just begun to sort things out. His initial confusion was now replaced by a bewildering recognition - a recognition of the audacious magnitude of what seemed to be taking place in his son's life. If he was asking for Dhruv, then there had to be something not quite right somewhere.

"You should be telling us that," said his father as he sat on the bed and ran his fingers through his son's smooth hair. "Abhishek told us that you rode pillion on Uncle Dhruv's bike, but when we spoke to Dhruv, he replied in the negative. So, where have you been all these days? Do you remember anything at all?"

Anil gave him a puzzled look. "Yes! Uncle Dhruv offered to drop me home. Abhishek was with me, but he declined because we were just outside his house."

Mrs. Sheen and Mr. Sheen looked at each other dumbfounded. The rudimentary sense of relief that they had experienced on seeing their son, gave way for an uneasy feeling. They tried to appear calm but failed. Something was inapt, and they had a vague feeling that the coming days would most likely not be very happy.

Mrs. Sheen was exasperated. "Oh no!" she moaned as she ran through her thoughts sieving her reservations and restlessness. She felt faint and dehydrated. She was desperately in need of some water. Mr. Sheen obliged by getting it for her. As she sipped the water slowly, she was abruptly stopped on her mental path by an awkward observation.

Acutely aware that the accretion of sleepless hours may have taken a toll on her tired eyes, she thumbed them to ensure that she was not merely imagining the unreal, and then looked at him again. No, she was not mistaken. Clutching her face between her two palms, she moaned in sheer frustration to Mr. Sheen, "Look at his eyes. I mean, his blinking. There's something grimly surreal about it. You had better call for the doctor immediately. How is it that the doctor at the police station did not notice this? It's so unmistakably evident."

Mr. Sheen responded by looking at his son seriously and intensely. "You are right," he said at length. "The blinking is indeed..." He kept searching for an appropriate word to describe it but finding none, all he said was, "it's curiously different. The doctor may not have noticed because he was unconscious at that time and his eyes were closed. He did insist, though, that I give him a call soon after he got into the waking state. I'll call him right away."

"No, don't. I mean don't call him. Call Dr. Sharath. He would be able to throw better light on this. After all, Anil has literally grown under his supervision."

"Yes," acquiesced her husband. "He would know better. I'll call him, but not now. It would be impolite to disturb him at this hour."

* * *

Dr. Sharath came across as a philanthropist who was in a tearing rush to assuage the world's ailments in a timeline that was too short. A popular figure in that locale, he enjoyed being needed and was characterised by his bag of emergency supplies which was an inseparable part of his robust personality. When he dropped by in the morning, he agreed with the Sheens that the blinking was indeed unnerving.

"Riveting!" said the startled doctor. "That's the word. It's so mechanical that I think I'm justified in presuming that it could be a cause for alarm, or maybe I'm wrong. The time gap between 2 successive blinks is too synchronized to perfection, unnaturally motorized and well timed for comfort. To put it more explicitly, it's preternatural. Since when has this been happening?"

"We noticed it very early this morning, soon after he regained consciousness. We intended to call you then but the Sun had not yet risen," stated Mr. Sheen

"What?" The doctor was stunned. "He was unconscious? For how long? And why was I not informed about it?"

"It's not what you think," clarified Mr. Sheen. "He had been lost for a few days. He was found insentient in the local garden, yesterday morning. We perceived this, and I'm referring to his blinking, only after he revived from a state of stupor late last night. It got us worrying. Do you think it will alleviate?"

"I see no reason why it shouldn't," replied the doctor. Then turning towards Anil, he asked, "Do you feel any pain?"

"No," replied Anil.

"Can you see clearly?"

"Yes, doctor."

"Any discomfort?"

"No."

"What's to be done?" intervened a distressed Mr. Sheen.

"Nothing for now," advised the doctor. "There's no redness, no swelling and no pain either. Besides, he doesn't look to be suffering from any discomfort because of it. Let's wait a day or two. For all you know the blinking may just wane. But yes, see to it that he rests with his eyes closed all through the day and night, and ensure that you keep cold compressors on his eyelids for comfort. I'll drop by tomorrow evening to have another look."

Much later, in his confidential report, Dr. Sharath made a noting in bold red letters. It was a remark which stated explicitly that the 'blinking' had no parallel in his recorded history of personal practice. It called for fact-finding on the subject.

The next evening, when Mr. Sheen drove in through the building gates and took a curve that led him onto the parking area, he was relieved to see his son playing with the other children. It had been an extremely busy day at the office, and had it not been for Dr. Sharath's impending visit, he would still have been at work.

"Anil seems to be in better shape now," he remarked to his wife as he loosened his tie and sat on the ottoman. "Thank God everything's returned to normal."

"Yes," replied Mrs. Sheen, "but his 'blinking' continues to bother me. I hope it gets better soon. Everybody's been commenting on it."

"Don't stress yourself unnecessarily. He seemed absolutely fine when I saw him down there." So saying, he gave her a quick glance. He noticed that she looked tired, testy and most unlike her garrulous self. "I hope nothing's wrong with you. You appear jaded. As I said earlier, just relax. Everything will fall into place, and things will return to what they always were."

"No, I'm not worried about Anil. I trust the doctor," replied Mrs. Sheen.

"Then why do you look so washed out?"

"It's my back," she replied. "It's aching. The continuous strain of the past few days has taken a toll on me. When is the doctor expected? I'll request him to prescribe something for immediate relief."

When the doctor showed himself later, he pronounced that Anil was as fit as a fiddle.

"What about his 'blinking'?" asked Mr. Sheen. "He's been getting undue attention because of it."

"His blinking?" replied the doctor. "It's too early for me to comment. It does not appear to have any adverse effect on him. Like I said last evening, let's wait a while and see if it subsides on its own. If not, I'm always there."

"And my wife, doctor?"

"Why? What's wrong with her?" asked the doctor.

"She's been complaining of languidness and a pain on her lower back," stated Mr. Sheen.

The doctor diverted his attention to Mrs. Sheen, "What's happening? Since when have you been feeling low? Buck up. Anil has been found. You should be happy."

"I am," she replied. "I'm happy my son's back, and that's not what's got me worrying. It's my back. It's paining and does not seem to be getting any better. Besides, I've been feeling abnormally tired since last evening. I don't generally feel so."

The doctor went through the motions of checking her pulse, her tongue and her heart beat. She appeared to be fine, except for her eyes. There was something in them that he could not quite define – fatigue or despondency or both. He deduced that her present physical condition was a result of the events of the past few days. He prescribed a few OTC medicines and advised total rest. He promised to drop by the next day to check on her health, but the morning that followed was far from optimistic for Mrs. Sheen. She was too worn out to even prepare breakfast. 3 days later, she died.

* * *

Flushed with the success of the second part of the Plan, the Martians discussed the next level of action with vigour. There had been no dubiety about the configuration of the chip, but the fact that it worked so seamlessly even with the subsequent sample, proved that their holstered calculations were totally foolproof. It reaffirmed that their sense of detail for implementation was without flaw or blunder. Their expectations scaled higher.

"A very smooth operation, despite its convolution. Besides, the results are quicker here," proclaimed Pontus. "Time to consolidate on our exito."

"Yes, but with discretion," cautioned Maxus. "The results are pleasing, but quick, and such quickness may cause ripples in the Venusian territory, if we are not careful. Their unsettling intervention is an unpredictable prediction. I know it was you who cautioned me when I asked for faster ramifications, but now, it is I who wish to remind you that it's never wise to disremember them. Their perspicacity can never be undermined."

"I'm tired of discussing them all the time," said an infuriated Pontus. "They seem to dominate every conversation we have. I've told you earlier and I say so again that I've been scrupulously cautious this time. It's been eons since our last confrontation with them; enough time has elapsed from then on. I would not hesitate to accept the likelihood of their laxity in the matter. If they had sniffed anything untoward so far, their righteous attitude would have elicited a response by now."

"Their lack of response should not be construed as reason enough, to take things for granted," warned Maxus.

"But we have executed a far better and satisfactory approach this time," countered Pontus. "Quick action! Utmost canniness! Absolute secrecy! And the results are there to see. It only encourages me to be faster in releasing more specimens to conclude our mission within a time frame that'll give these nosey parkers very little time to put a brake on our blueprint, if and when they smell something fishy."

"One never knows about their interference," enunciated Maxus, "but to your attitude, I can only say 'Thumbs Up!' Our glorious planet, come what may, must remain unchallenged until the end of Time."

*** *** ***

CHAPTER VII

It was a day like any other at the Institute of Research for Space and Natural Sciences. Busy. Rishi jerked himself from his seat, arched his back and walked towards the vending machine. He was in need of a cup of coffee, before immersing himself into his work again. Alex joined him. "Heard the latest?"

"Latest? On what?" Rishi countered.

"On dying females!" replied Alex

"Referring to Likwi, by any chance?" queried Rishi. "The place of doom for females! Is that still happening? Been too occupied with other things to keep tabs on that. Besides, the 'Team' we had deputed, reported nothing of significance."

"Yes, it's still happening," asserted Alex. "But my reference is not to that place, but to a similar phenomenon that has been noticed elsewhere."

Rishi held his half empty cup in mid air and looked at Alex incredulously. "Not joking, are you?"

"No. I'm earnest," stressed Alex. "Twenty have already died in a span of just 5 days, and they are still counting and," he hesitated before proceeding, "as in the earlier case, all of them are females."

"Pretty unsettling!" Rishi appeared downhearted.

"Not pretty but very," emphasized Alex. "Something direful is latent in these manifestations. Nowhere in the history of mankind has such a thing been recorded. That it should happen at all is astonishing. It needs a tourniquet to staunch the haemorrhage of deaths. Why don't we depute our team there, too?"

Alex sounded intense and determined. He was not the kind to humour himself with wasteful exhortations. His punctilious brainpower did not permit him to do anything extravagant, and so when he did make suggestions, which were really not many, they deserved reflection. Rishi knew he was right in saying that there was something sinister underscoring the Mephistophelian events.

"I agree," Rishi acceded, "I think we should chase this."

"Was there anything conclusive in the findings of the 'Team' that visited 'Likwi'? Anything creditable?" asked Alex.

"Like I said earlier, nothing noteworthy," replied Rishi. "No trace of a disease or epidemic. Nothing malapropos. Even the most qualified of the doctors were unable to place their finger on to the root cause of this malady. The only thing that created a flutter was a lad called 'Gattu.' I believe he blinks funnily."

"Blinks funnily?" asked Alex intriguingly.

"Yes. He keeps blinking mechanically 3 times in succession. Apparently, the blinking is almost akin to a keyed doll that blinks when the key is triggered," stated Rishi.

145

"Interesting!" exclaimed Alex. "Was that also indicated in the report?"

"Yes," said Rishi decidedly. "It was imperative to record even the most frivolous thing. You never know when it may prove to be of some significance in the future."

"You're right. Never treat anything as trivial," agreed Alex. "The most belittled, sometimes, assumes the greatest importance. Anything else?

"No... wait a minute," shilly shallied Rishi. "There's something else that deserves a mention. Diane, who was a part of the team, complained of a minor backache soon after the team returned from there."

"What's so unusual about that?" commented Alex. "Females are all the same. If they are not complaining about minor aches and pains, they keep carping on the discomfort of expeditions like these. Our attention is always diverted to their well-being instead of being totally centred on the Project in hand. Not that I am biased, but let's be practical. I don't deny that they are on par with the males where their intellectual inclinations are concerned, but personally I feel that they are biologically more suited for corporate affairs. That's why my insistence that no female should be included in the team on an outdoor project. They have neither the vigour nor the stamina of a human male."

"Uff!" Rishi let go of a sigh. "That was a long discourse. But I beg to differ. Diane is as tough as nails. In fact, she has always been a part of our team and rarely ever whines. Have you seen her reports? Admirable!

She sure knows how to distinguish between fact and falsehood. The report on Likwi was no different. This was the first time ever that she had complained, and that too in respect to something she thought was either totally irrelevant or totally relevant to the context. And by the way, this has not been reported in the 'LIKWI' dossier. There's no written statement on it."

Alex looked up with a jolt, "What has not been reported?"

"The backpain experienced by Diane," answered Rishi.

"Why not?" queried Alex

"She thought that such personal remarks were unnecessary in a very compact professional report," continued Rishi.

"Oh!" responded Alex.

"But now when I think of it, I think we should have included that piece of information too," said Rishi.

"Why the afterthought?" asked Alex.

"She's got a great eye for details," explained Rishi. "Has always played an elemental role in getting to the bottom of most of the cases that we had managed to solve in the past few years. Her backpain was just a passing statement she had made during my conversation with her. At that time, I dismissed it carelessly; I thought the overemphasis she put on her pain appeared to be rather disproportionate to an otherwise sedate account... but come to think of it ... the more I dwell on it, the more I feel that it may have some significance... anyway let's forget it... maybe it's just my imagination running wild ..."

"Now that the subject has cropped up, you may as well discuss it. Let me be the judge," probed Alex.

"All right, if you say so," assented Rishi. "She remarked that when she looked at this boy 'Gattu' she felt disconcerted. Like everybody else, she too noticed that he blinked thrice in succession continuously and mechanically."

"But we already know that, don't we? It has been recorded too. That's not new," commented Alex.

"That's not what's new," allowed Rishi. "What's new and not recorded, is the fact that she noticed something that others didn't. The intensity of his glance, though innocuous, was strangely quite threatening. It created a patch of pain on her lower back. Initially she dismissed it as mere coincidence and continued her dialogue with the local officials. But her curiosity got the better of her. She arranged for a special meeting merely so that she could look at him again, deliberately, just to ensure that the affliction was for real. She knew she was not mistaken when she felt it again. She dared not look at him then on."

Alex responded by dilating his eyes. "Go on. Sounds like a story I read when I was a kid."

"It's no laughing matter," admonished Rishi. "When she enquired of the rest of the team members to ascertain if anyone else had experienced something similar, they all replied with a big 'NO.' Her feminine instincts surfaced, and she concluded that it would be unsafe to meet him again. Consequently, she avoided him for the rest of the tour. She felt her pain would worsen every time she would lay her eyes on him. Ever since, she has been feeling a little lethargic. In fact, she's on some healing therapy, now."

"You feel the incident requires further reflection?" enquired Alex.

"Not sure," said Rishi. "It's just a thought I have had at the back of my mind. Do you realise she was the only female member of the team, and she was the only one who sensed it? The others were all males. These figures of death that we keep getting to hear of, are all females... I wonder..."

Alex summarised Rishi's suspicions and conclusions. "I guess there's some truth in what you say. The sooner we tail this, the better it'll be. It's a contingency that calls for the highest level of attention."

Rishi who had just finished his third cup of coffee nodded his agreement to affirm that he had made up his mind.

"Should we include Diane too in the team for Hosuru?" asked Alex. "Her presence would certainly be useful, but I wonder if she'd agree considering her last experience. It would be interesting to see what her 'eye for detail' manages to glean."

***

The newsreader was devoid of expression as she read out the report in a sombre tone,

"... And now for some breaking news that is taking the world by storm; it was 'Likwi' and now it is Hosuru which is grabbing the headlines for a similar reason. At the last count, the number of females dying here for no apparent reason has already risen to 50 in a span of merely 7 days.

Most of these deaths have occurred in a school where the victims are teenagers between the ages of 13 and 16. A few stray cases of older women have also been reported. What is alarming is that all of them are females. A pattern, like that in 'Likwi,' seems to be emerging causing grave concern to the authorities. A team which had earlier visited Likwi has already been despatched by the Government of the day sans females. The Government is fighting shy of sending females because the two female members of the earlier team succumbed to death immediately on their return.

Analysts are trying to establish a connection between the two. The Institute of Research for Space and Natural Sciences, i.e., IRSNS had also sent a team to Likwi but failed to find anything of significance. They have yet to confirm if they intend to send their team to Hosuru, to help with the inquisition. We offer our condolences to the bereaved families. Continue to remain tuned to our Channel while we keep you posted on the latest that is happening on this front."

Rishi and Alex exchanged meaningful glances. "Amazing! That says it all. No females. The authorities say that, and so we say too. No Diane. Certainly not, after having heard this piece of information. The decision has been made for us," said Alex.

Rishi smiled, "Let's see."

Diane entered just then. "Heard the most recent? Boy! Was I lucky? A longer wait at Likwi and guess what? No conversation with the two of you. Lady Luck does seem to have a hand in this thing called 'Life.'"

"How scary!" Rishi responded. Alex agreed.

"Heartrending, isn't it?" Diane continued on a sombre note. "At Likwi, we dismissed the deaths summarily quoting lack of modern medical facilities. The local doctor had been dogmatic in his contradiction of our conclusions. I still recall his mesmerizing voice telling us succinctly, "Likwi may be renowned for its isolation rather than for any other distinction, but as far as medical facilities are concerned, we are on par with the rest of the world. I have ensured the best treatment to my patients."

"That's impressive," replied Alex.

"Yes, isn't it?" said Diane. "Apparently, he was a star student and his passion for philanthropic causes guided him to practice in that non evocative town or rather village."

"Kudos to him," applauded Rishi. "But he is not our subject of discussion. You are."

"I?" asked Diane mildly surprised.

"Yes. We were discussing the probability and possibility of sending a team to Hosuru and were toying with the idea of including you. The news has aided us in reaching a conclusion. The way we see it, you are, in no way, going to be a part," concluded Rishi decisively.

The vociferous person that she was, Diane experienced a battle of wits between her heart and head. Her head, as usual, won.

"Oh, shut up!" She snapped. "You should know better than to make gender-biased decisions. You know what you have done? Provoked me to defiance."

"Hey," mocked Rishi, "don't inculpate us. It's not our arbitrament. The authorities have decided. Undeniably, they have your best interests at heart. You can't disregard them."

"Best interests?" mocked Diane.

"Yes, it's a precautionary measure. The idea is to safeguard females," stated Rishi in a matter-of-fact manner.

"I still insist on accompanying the Team," persisted Diane.

"In a rebellious mood to defy orders, aren't you?" asked Alex with uncharacteristic savageness.

"Why not?" Diane was suitably grave. "Females dying; more reason for a female to give her point of view."

Rishi looked at her appreciatively. She may well have been discussing a trip to Paris instead of a place where her life could be at stake.

"But what about the backache you suffered from, the last time you visited Likwi?" Rishi asked, at length.

"Ah yes!" replied Diane. "That pain! It continues to recur. Gets better only when I close my eyes and rest awhile. Or possibly it's the meditation. I am strongly into it these days. Steve tells me that the ancient Indian sages have stated that there is only one nonpareil way to free ourselves from this life of misery. That is meditation. It gives us good health and it gives us good 'peace.'"

"Looks like your visit to Likwi has had a vitalising influence on you. It appears to have altered your perception on life." said Rishi.

"Not my visit to Likwi, but this backache of mine," answered Diane. "Ever since Steve introduced me to meditation to relieve me of my pain, I have unfailingly dedicated some time to this spiritual contemplation. He says that anybody and everybody who meditates in a technically perfect way would receive all the benefits of good health. He even tutors me on doing it in a scientific manner. I'm feeling a difference already. The ache still persists, but it has minimised."

"Are we to presume then, that you are serious about your decision to visit Hosuru with the team?" asked Alex.

"Absolutely!"

"In any case we shall be obliged to include you," said Rishi, "not because you are a female but because you are the best. Hopefully, your inclusion will bring a material alteration to our report, and offer a new insight to a problem that appears to threaten the very fabric of our existence. A few legal details may have to be complied with. A word of caution, though. Take care. Should you perceive even the slightest hint of danger to your life, do a quick U turn. The others can take over from where you have left."

Diane sneered, but was grateful for the concern.

"But before you leave, you will have to sign an all important legal document; it's imperative," stated Rishi determinedly.

"Legal?" asked a visibly surprised Diane.

"Yes," said Rishi. "Your signature has to be affixed on an affidavit claiming that you are doing it of your own free will and would absolve anybody or everybody of all untoward ill-starred happenings, if any."

"Is that de rigueur?" Diane asked.

"Yes," Rishi replied. "It has been made mandatory by the Government after they lost 2 of their female members who were a part of the investigation team in Likwi. A circular will be sent to that effect to all private agencies."

Diane laughed a laugh that freshened the atmosphere otherwise laden with negativity, "Does that leave me with an option?" She loved accepting challenges. It gave her a feeling of usefulness.

"Let's set the agenda for the trip then, and you better be quick with your preparations. You may have to leave sooner than you think. A visit to the school is a 'must'. That's where maximum deaths have occurred, remember?" reminded Rishi.

"Yes. That's a priority," agreed Diane. "We will probably find some vital clues there. I'll be there, while the rest of the team gets busy elsewhere."

"Rightly said. It's germane to the context. That's the place that may throw up something conclusive... something that will facilitate us to make headway in the right direction," concluded Rishi.

"Amen," said Alex.

* * *

Diane reached the school a little earlier than the appointed hour. The Principal, a grey-haired man, was debonair with small precise movements. His bland face appeared expressionless except for a constant twitch of nerves on the side of his right temple, while his figure was so plump that even a tailoring of genius could do little to improve it. His faintly accented voice, of late, resonated with a sense of worthlessness, tension and failure. The reputation of his school was at stake for reasons that were anything but academic.

It was unbelievable that, despite the advances made in medical science, the best of doctors pleaded nescience to the cause of death of the young girls in his school. The media had been merciless and the parents furious; understandably so. He himself was clueless, and had no justification to offer to the gracious lady with whom he had an appointment that morning. He was wondering on the remedy that needed to be adopted to avoid further damage, when Diane knocked on his door.

"Come in," he said, smiling a ghost of a smile out of sheer politeness.

"Good morning," greeted Diane, beaming at him in a fashion that doubled itself with an introduction. She gauged him to be in his early fifties.

"Good morning," acknowledged the Principal. "So, you are from the 'Institute,' and you think you would be able to be of some assistance in identifying the problem." He spoke with the clarity and firmness peculiar to men who knew their job and were accustomed to giving orders.

"Yes! Although if you do ask me how, I would have no concrete answer to give you ... at least not now." She gave an ill-concealed careless shrug.

"The doctors say the same too. I'm still waiting for them to reach a conclusion - positive or otherwise," pronounced the Principal.

"We are aware," stated Diane blandly. "But we are not doctors and we look through the phantasmagoria contrarily. We have been wondering if there is another angle to it."

"Another angle?" The Principal appeared confused.

"Yes!" underlined Diane. "But that can be discussed later. I came here to meet you specifically because it is rumoured that parents are planning to deregister their young daughters from your school fearing impending deaths. Is it true?"

"Yes! That's true," conceded the Principal. "And I can understand why. As a parent, it's natural to think on such lines. It's incredible that only last year we were awarded and felicitated for being one of the finest educational institutions in this part of the world. And this year! This year's been a nightmare. I'm almost beginning to believe in necromancy and witchcraft. Keep wondering if it's a ploy by one of our rivals to undermine our reputation."

"Quite possible," replied Diane not because she agreed with him but merely to assuage his hurt pride. "But conclusions are baseless unless substantiated with evidence. And for this, I would need your total cooperation."

The Principal wore an emotionless mien on his face while lending a patient ear to her. But beneath that placid countenance he felt an emotion he could not specifically identify - hope, perhaps?

"Of course!" he declared. "I am willing to do anything. I would voluntarily cooperate in every which way you think would be useful, if I am assured that my school will be exonerated of all the cruel baseless accusations that are being hurled at it. I need all the help I can garner in these powerless moments."

"Baseless accusations?" decreed Diane. "They may seem baseless but they are irrefutably indefensible. You can't deny that with so many sudden dubious deaths occurring in such an ephemeral period, the allegations, at least at this juncture, appear justified."

Her stint in the University, years ago, had taught her that the only effective way to impel men into action was fear, and she hoped to do that by making a statement that would coerce him to cooperate without concealing crucial details. The Principal appeared visibly perturbed by that statement.

"I'm at a loss to give a bona fide reason for the sudden turn of events," he uttered waveringly. "The unfortunate deaths are an inexplicable shock to me. I've been tossing around thoughts and reasons with my staff members, only to find myself running round in circles. As an outsider, maybe you can lend a helping hand. What do you think needs to be done to avert it?"

"That's why I am here," replied Diane, "to see what we can do. I have been told that the deaths have now subsided. That should give you some respite."

"True," acquiesced the Principal in a voice that was low but coupled with a breathless quality of defencelessness, "but I need to be assured that this lull in deaths is not temporary. The deaths have to cease completely, never to reoccur again in the future. Have you anything specific in mind?"

"I would like to start with personal visits to the classrooms," she averred.

"How would that help?" asked the Principal.

"I'm not certain. But I need to begin somewhere."

When Diane had expressed her desire to visit the school, her resolve had been crystal clear, but now as she looked across at the man seated opposite her with an expression of cleverly concealed incredulous horror, she realised that what was overriding was not just clarity of purpose but a motion of purpose, no matter how small - a sense of activity going step by step to a goal that was ambitiously set by her, and for her. She smiled a faint smile of sadness. Steve was right. There was greater pleasure in solving humanitarian problems than in being preoccupied with self-gratification. As she got up briskly from her seat to accompany the Principal on his round of the school, she gave the impression of a woman who was on the verge of finding a solution that was not too far away.

"Shall we?" She asserted hoping to inspire the Principal's declining confidence from degenerating further.

"You're sure?"

"Yes. Like I said a little earlier, I must start somewhere, and this is the only place that I can think of at the moment. My action may defy logic, but not my instinct."

"Well! As you say."

The visits to the classrooms, regretfully, did not yield the lead expected by her. Perhaps she had been wrong in adjudging that she could just walk in and identify the boy. Maybe she needed to be more forthright.

"Nothing irregular," she remarked in a tone that neither expected nor permitted an answer, as she seated herself in the Principal's cabin for a break, before resuming on her next set of rounds.

As she sipped the coffee slowly, she conversed with him on random subjects, until she thought it would be appropriate to deflect her confab to a subject that needed immediate consideration. With a faint solemn stress in the tone of her voice she said, "I recollect listening to some information on Television about a boy mysteriously disappearing and reappearing after a few days. Apparently, he belongs to this school."

"You are right. He is a student of Grade X."

"What's his name?"

"Anil. The poor child lost his mother immediately after that."

Stunned, Diane remained still, trying to grasp the parallel in the 2 ordeals - the one at 'Likwi' and the other here at 'Hosuru.' Jerking her head up she said, "Now, that's mindboggling! I shouldn't be telling you this, as it may have no bearing on this case, but there does seem to be a great similarity to another incident which occurred in another place, a few weeks ago. There too, a boy mysteriously disappeared and reappeared after a few days. He lost his mother soon after."

She paused, and looking directly into the Principal's eyes asked, "Have we visited his class?"

"Not yet," said the Principal. "I would counsel you to be discrete when we do so. It is imperative to avoid paying too much attention to him. The poor child has been subjected to a lot of tribulation already."

And as an afterthought he added, "He is easy of recognition."

"Easy of recognition?" Diane was stupefied.

"Yes," affirmed the Principal. "He is the only one in the school who moves around in dark glasses throughout the day. Has been advised to do so by his doctors."

"Dark glasses? Why?"

"Ever since he was lost and then found, he appears to have some difficulty with his eyes."

"Difficulty?"

"Yes. Keeps blinking in a funny manner, you know."

"Funny manner?"

"Blinks consecutively 3 times in a very mechanical way. Like a robot, you could say."

"What a coincidence! The boy I referred to earlier, also does that."

"Really? That's amazing. You should discuss it with the doctors here. They find his case an enigma."

"Permit me to meet the boy first."

"I will. On second thoughts, it would be advisable to meet him in my office rather than in his classroom."

"That would be perfect," Diane replied with eagerness. Her thoughts were in turmoil. There was a possibility, remote as it seemed to her, that there may be some connection between the 2 paradigms. Gattu then... and now Anil.

Anil was summoned to the Principal's office. He was awkwardly tall for his age. Diane looked at him silently for a long moment with a certain hesitancy wondering if he could read her thoughts through those dark glasses that appeared uniquely different from the ones in the retail stores.

"So, Anil, your Principal tells me that you are one of the most intelligent students in the school," she addressed him.

"Thank you, Ma'am," replied Anil as he shifted his balance from one leg to another with a youthful pride, eager for acknowledgement.

"Those glasses that you are sporting - do you wear them all the time?"

"Yes. The doctors have advised me to do so."

"They seem to be distinct from the normal ones I generally see in the shops."

"This pair has been tailor-made exclusively for me."

"Really? Why?"

"Don't know."

"Do your eyes pain you?"

"No, they don't. I told the doctors I feel better without them. But they insist that I sport them all the time except while sleeping."

"Have they told you why?"

"To protect my eyes from the bright sunlight and the polluted atmosphere. They feel that my exposure to these factors may worsen my blinking."

"Is that so? May I have a look?" Diane's head dropped in assent, but her hand reached out for the spectacles.

"No Ma'am." Anil took a step backwards. "The doctor has specifically told me not to look at any one without them."

She inclined her head and said persuasively, "Only for a short while."

The Principal intervened. "Anil, I think you can remove your glasses for a moment. Ma'am is curious because she knows of someone else who blinks just the way you do. She merely wants to confirm that his blinking is like yours."

"All right, Sir," said Anil, temporarily lifting his eye accessories and placing them on his head.

Diane was startled and without knowing what prompted her, heard her own voice rudely ordering him to put them back again quickly. "You may go now, Anil. Thank you." The 'blinking' was familiar... achingly familiar and sinistrous. She felt a similar stab at the same spot at the lower back as she had felt when she had first set her eyes on Gattu... and that strange ennui... this blinking... would she die like the others? Prehensile... that was how her mind worked. She was gifted with the art of seizing bits and pieces of seemingly unconnected situations and putting them in relevant slots to reach a logical conclusion. Here things were slightly different. She could see the connection as clearly as she would have seen her reflection in a clear pool of spring water but could offer no rational explanation for the same. But one thing was for sure... the 'blinking'... it had a hair-raising effect on her.

Soon after Anil left the Principal's cabin, she requested the Principal for a glass of water. She was aware that he was looking at her furtively fathoming her reaction which, though not too blatant, was not clearly hidden either. It was a look that crawled searchingly through her, silently questioning the effect the boy who had just left his cabin had on her. Perhaps his action was justified as she had only partially succeeded in concealing her torment.

The insupportable pain that she had experienced while looking into Anil's eyes, as in the case of Gattu, was not mere physical exhaustion but a sensation that was hard to endure. She felt her anxiety radiating from her like waves of heat. Standing up she swivelled her hips. Yes the pain was no different either, only it was more intense than before. For a moment, time stood still and her mind ran in circles. She sought the Principal's permission to rest a while to regain her composure before continuing her conversation with him.

"Yes, Anil's blinking resembles the other boy's," Diane declared. "The coincidence is startling, to say the least. Both the boys blink very much the same way, and the areas where they are found seem to be witnessing female deaths at an alarming rate."

"You think you are on to something. Do you feel this child is in some way related to these deaths?" The Principal was aghast.

"I think so, although I am not too sure. Did the deaths start subsiding after Anil started wearing these weird eye accessories?" asked Diane.

"Now that you say so, perhaps yes. But I haven't really given much thought to that."

"Think it over," asserted Diane. "I think I've found what I've wanted to. By the way, I would request you to keep this meeting confidential. What has taken place in this room must never go beyond its walls. It's important to get to the bottom of this as soon as possible. If my intuition serves me right, I think I have found a clue; a vague one, to be sure, but it'll probably lead us on to the right path. Thanks a lot. I'll take your leave now."

"Shall I call for some more coffee?" asked the Principal politely.

"No," replied Diane, "that would be unnecessary. I would like to have another glass of water, though. I would have left immediately, but this backache of mine stalled my exit."

She rose to leave. The Principal accompanied her to the door, bowed courteously and remarked, "I hope to receive some reassuring news from your end soon."

Later, on reflection, the Principal found himself surprisingly fuzzy with the details. The conversation with the lady and what ensued before, after and in between was so overwhelming that it defied lucidity.

* * *

CHAPTER VIII

Athenia looked disbelievingly at the electronic 'Earth graph' that was designed to document every death with the specifics. It was the key. Her discerning mind refused to believe what her eyes saw. She gave a second look, checking and revisiting the dots singularly and severally. No, she was not mistaken. It revealed an abnormally disconsolate picture. The inconsistency was uncharacteristic - an obvious pointer to a peculiarity. The last line in the bar graph had dematerialised. Her uncertainty shaped into confirmation.

She looked at Phoebia with her head bent and an apprehensive consternation, "This is disturbing. There's a rapid and dramatic decline in the number of females. It defies convention. Do you recall the discussion we had some time back, on the subject? It had been a constructive exchange of views. You had advised me to give it a more attentive look and report any aberration. I've been doing precisely that. The observations, to say the least, are very agonising."

"It would be best if you would be more explicit. I'm in no mood for riddles," said Phoebia.

"Take a look at this bar graph," verbalized Athenia. "There's a striking reorientation in the male to female ratio. The last line has disappeared."

Athenia and Phoebia stood shoulder to shoulder inspecting the electromagnetic graph right down to its most microscopic details. To the naïve it seemed to be an ingenuous portrayal of facts, but to the initiated the barely discernible fluctuations were a cause for alarm.

As the red and blue dots kept flickering Phoebia was perforced to comment, "Oh no! This is extraordinary! Are you sure there is no error? The programme is extremely hypersensitive - even the slightest erratum can throw it completely out of gear."

"No, no. That can never be," stated Athenia intensely. "It's our latest brainchild, and you know how fastidious we are about ensuring that it never crashes. But if it satisfies you, I'll rerun the gizmo and interface the programme to relate it to the cue-numbered dots."

At length she darted her eyes towards Phoebia. "It's established. The repercussions don't make for a reposed state of mind. This is what happens when we take too much for granted. A woebegone condition!"

"We were never laid-back," insisted Phoebia.

"But we had diverted our engrossment towards Mercury, for a while," countered Athenia.

"The planet deserves as much consideration," justified Phoebia.

"I don't deny that; but prioritizing takes precedence; or else, devils rush in where angels fear to tread," declared Athenia.

"And who's the devil here?" asked Phoebia.

"The Martians," highlighted Athenia.

Phoebia looked at her quizzically, "That's a bold allegation. Unconfirmed and harsh until proven otherwise. I'm more inclined to give the title to the Earthlings. They can't be entirely disregarded. They've done a spiffing job of infecting their own world with their besmirched brand of politics and philosophy." She made no attempt to disguise her contempt or bitterness.

"What prompts you to say that?" enquired Athenia.

"Go back to a time when something akin to this had occurred," reminded Phoebia. "We had been foolish and hasty in laying the blame at the door of the Martians. Our conjectures proved to be unfounded, and we were forced to put our tails between our legs. Abortion of female foeti, female infanticide, dowry deaths and brutal rapes, had all led towards a reduction in the number of females; and who was to be inculpated - the Earthlings."

"That was an impermanent misfortune; a mosaic of wretched events," defended Athenia. "It has not escaped from my memory. But the crisis remedied itself as quickly as it eventuated. Don't forget that there are good Samaritans too, there. When the flagitious cause damage, the venerable take centre stage to remedy the misadventure. But now, the pattern is distinctly different. It is symmetrical... alluding to an arithmetical precision - definite, repetitive and yet quite unplaceable. I don't think I'm entirely wrong in considering a Martian intercession. Don't forget, there's a precedent. They have been guilty of such a misdemeanour in the past."

"Only once before," emphasized Phoebia, "and hopefully that would have been the last. Remember, they were warned. Besides, the Martians are not the only ones with a predilection to conquer and rule. The Cosmos has countless such empyrean units. It would be brazen to base our qualms on unconfirmed contentions. Why do we always have to zero in on the Martians? How well do we actually know them?"

"Well enough to grasp their unyielding desire to annexe cosmic power at any cost," enunciated Athenia. "Their attitude only serves to exacerbate my suspicions. None but we know the magnitude of the violent Martians. They will bomb galaxies, destroy planets and while at it, we may be wiped out too. Their desperation to conquer Space has never been a secret. Their first target is Earth, a planet that finds itself in an extraordinary position."

"A position that's fortuitous," remarked Phoebia. "It is unfortunate that the Earthlings take all their endowments for granted."

"True," agreed Athenia, "and the Martians, like us, know that. It would have been easier for them to bombard it but, then, that would lead to dissolution of matter, resulting in its total extinction from the Cosmos - an unappealing idea."

Phoebia concurred.

Athenia continued with her rationalization, "They desire the planet minus the human species. And what better way to do it, than this. Decimate the females, orphan the males and eventually wipe out the entire human race... a sublime, or should I say a heroic achievement for them. Yes, I'm sure, I'm right in hazarding that they are responsible. I have never been completely sold on their ostentatious display of 'peace.' The niggling doubt has always existed and not for an instant will I ever discount their drive to conquer Earth. It's their foremost goal, and it's unlikely that they will ever abandon it."

Athenia's raisonnement made sense to Phoebia. Once before, they had been forced into retaliation, when the Martians crossed their boundaries, but that was no reason to hurl accusations at them all the time. Prejudices, preconceptions and ifs had to be relegated aside if an ad rem panacea had to be found within the quagmire.

"Don't get emotional. Once too often, you have let this feminine instinct of yours take over the rational side with dreadful consequences," Phoebia cautioned.

"No. I have been alert, all along," claimed Athenia. "The 'Scan Scope' has thrown up nothing of signification pertaining to Earth. No floods, no earthquakes, no hurricanes, no landslides, no meteoric entries... no natural calamities ... none whatsoever."

"Are you sure?" prompted Phoebia.

"Doubly sure," confirmed Athenia. "And perchance that had been the case, the casualties would have included both males and females. Why the bias for females? I have been spending unreasonably long periods trying to solve this mathematical equation that is presumably impossible, and one which involves both the males and the females. One glaring feature that makes me uncomfortable is the observation that the stylus moves back and forth endlessly on the female gender only. The males are nowhere close in the frame; only females. I repeat, only females!"

There was a strong imperative undertone in her words that riveted Phoebia's attention.

"I've heard you loud and clear. You are justified in asking questions and listing your reservations but that still does not legitimize your innuendo, even if the targets are exclusively female," answered Phoebia in a soft voice. "But I appreciate your paraenesis to resolve the issue. What do you think needs to be done?"

"Play the detective," replied Athenia. Her voice sounded unusually determined. "A solution can be reached only when we find out why it is happening and what the cause of it is. The task is intimidating because there is very little to go on, except for the numbers which are disturbing and unusual. Besides, as you rightly said, we can't overlook the fact that the Earthlings themselves are not favourably inclined towards females. This means that I will have to factor them too in my field of suspicion."

"It pains me to include the Earthlings in our list of suspects, but we'll have to. Their dislike for females is not concealed," stated Phoebia. "Their repugnance for them is beyond my comprehension."

"The entire humanity isn't averse to them," corrected Athenia. "In specific pockets they are glorified in excess. They even refer to Earth as

'Mother Earth.' But there are locations where they are unfavourably looked upon. In such places they are anaesthetized to suit the prickling concerns of the dominant males who mask their innate insecurity by subjecting them to obscurity. The males suffer from a delusion that their lineage is carried forward by them alone. Some even go as far as to abort a female foetus in the womb itself lest they are referred to as feckless. How cruel!"

"Inconceivable! What folly! And what is this lineage stuff? Another of their hare-brained acts?" Phoebia asked.

"Ha! Ha! Ha!" laughed Athenia. "Like I told you they are still in the primitive stage of evolution... the basic animal instincts continue to prevail in them... generations to pass by, before they reach the highest rung of the ladder of evolvement and the genetic code transforms itself to the acme of understanding that both males and females are required if they are to understand the 'Theory of Evolution,' better."

"Every reason to suspect them," reasoned out Phoebia.

Athenia disagreed. "No! No! This time, I can say with conviction that they are not responsible. At least, as of now, I don't suspect them. There's something more to it than meets the eye."

She spoke on a single-level tone, but her voice had the sound of efficiency making it hard for Phoebia to ignore.

"The swiftness of deaths," continued Athenia, "is causing grave concern; the decline was far slower the last time. Anyway, "who" is not of importance here, "why" is. A quick action is paramount if it must stop at the speed at which it has begun. Our involvement at this juncture would be beneficial for the entire Space. The consequences, if any, will be less incendiary than if we let matters stand as they are."

"Yes," concurred Phoebia. "We have to rectify the situation, or else, we may never forgive ourselves for our remissness," responded Phoebia with a faint hostility in her voice. "And now that you have aroused the third dimension of my intellect, my needle of suspicion does point towards the Martians. This brand of ruthlessness is so much a part of their psyche. But questions still lurk in my mind. It would be prudent to remember that no one's guilty unless proved otherwise."

Phoebia paused, a significant pause. "Get cracking."

"I will," asserted Athenia. "Let's start by rocketing our 'MAX VEN SPY' into Space straightaway to track unusual movements. A game plan can then be devised based on the data furnished by it. I'm neither in a mood for another 'Cosmic War' nor for holding ourselves responsible for perpetuating these very monsters who are out to extinguish the finest creatures in Space. Earth needs the arrant protection of an ally, if it must survive in its present form. We can't let a malefactor wreck the cosmic ecosystem."

Athenia returned her concentration to the graph and began reading and connecting the dots with such celerity, that they often appeared illegible. She was in haste. She didn't want to lose her assumption while she was at it. There was something about wrongdoers that aroused an instant animosity.

"Patiently! Calmly! We don't want to drop the scent," cautioned Phoebia. Athenia repressed a tart rejoinder. They had always worked together and trusted each other implicitly. Phoebia was the strategist while Athenia did the actual ground work. But unlike Phoebia, Athenia got affected by feelings, resulting in clouding her performance, more often than not. And she couldn't afford to do that now. They were no longer working on mere matters of Cosmic Space but the balance of the vast 3 dimensional region including interplanetary, interstellar, inter galactic and more importantly the life and death of the human species.

* * *

The report chronicled and tabled by the Team that returned from Hosuru had nothing irrefutable to say except that this was yet another freak case of nature that would be nigh impossible for ordinary mortals to solve. It was no different from the earlier one at 'Likwi.' The only dissimilarity lay in the date and the place. Rishi ran through it quickly... deaths... only females, symptoms... backache and exhaustion, prognosis... negative, striking observation... local boy missing and returning with blinking eyes... He paused and highlighted this in red with his marker. It was imperative to discuss the last point with Diane the next day. She had sought permission to stay absent that day.

Alex sipped his coffee at leisure and looked questioningly at Rishi who returned from the coffee counter with yet another cup in hand.

"Anything worth the connotation this time?" Alex asked.

"Not from the team, but from Diane, yes," Rishi responded. "She has come back with another of those acute backaches. She conveyed to me that if she does not rest right away, she may die of exhaustion. She'll show up tomorrow, which means, we'll have to wait until then to hear her. I am excited. She believes we could be on an interesting trail. Hopefully, we'll be more educated after a didactic korero in the morrow."

"Does she suffer from some chronic problem?" inquired Alex. "Haven't her aches and pains become recurrent?"

"This is only the second time she has complained of it," answered Rishi. "I don't remember her ever whining on health issues. If she has any, I'm not aware of it."

"No, I've never had any major health problems," stated Diane emphatically, as she recounted her personal encounter, in an inclusive meeting in Rishi's cabin, the next day, immediately after a routine gathering of the members of the Team. "This is the second time that I experienced it. The only difference was in the intensity. It was more agonising this time and I'm referring to both the pain and the fatigue."

Diane continued with her account, "There's a marked similarity between the 2 cases. It's a metaphor for an unseen hand. Now, don't ask me to qualify my percipience with any substantive reasoning. It would be problematical for me to do so. I base my surmise on 2 fundamental artefacts... the automated blinking and the 'absolute' that targets only females. It is tangential and disregards cognition."

"Sounds like fantasy. Absolutely implausible!" said Alex.

"And dubiously inconceivable!" added Rishi.

"I accede," replied Diane. She smiled with the wisdom of a woman who had just escaped death by the skin of her teeth. "Enough is enough. No more of this."

"No more of what?" queried Alex.

"This disconcerting rendezvous with 'blinking' boys," replied Diane. "I am not inclined to go through something comparable again. The pain I endured when I looked into the child's blinkers continues to haunt me; it was too excruciating. The Principal's face was a sight to behold when I asked the boy to leave the cabin rather gruffly and closed my eyes in meditation. He was dismayed. On another occasion he would have probably rebuked me indulgently for my lack of etiquette."

Diane's misgivings were nakedly revealed in her expressions. It was obvious that she was in no mood to subject herself to another comparable contretemps of this sort.

Both Alex and Rishi listened with rapt attention as she continued with her recounting.

"The searing pain, as in the previous encounter, was so momentary and fleeting that for a fraction of a second I suspected if I had actually felt it, but the corollary of physical exhaustion that followed was real - unreservedly real... a concrete proof that I was not in a dream. So knackered was I to do anything, thereafter, that I requested the Principal to let me rest awhile before bidding adios to him. It had been the same after my meeting with Gattu. If I do meet any of the two again, I swear, death will be the only consequence. Women ought to be warned against them. They are innocuously dangerous."

"You are being unfairly prejudiced and judgemental," declared Alex.

"No. I am not," stressed Diane. "I'm sure I'm not mistaken in my summing-up that these boys may have met all of the women who died, sometime or the other; and that too not once or twice, but several times. Like I stated earlier, if you ask me to validate my opinion with tangible evidence, all I would do, is put my hands up. It's just that everything adds up. Collecting data with the 'blinking' as the core guiding factor, is vital."

"You see a connection?" enquired Alex.

"Yes," underscored Diane firmly, "and a strong one at that. Link the likeness in the 2 cases..... Both were lost for a few days. Both returned with the malaise of awkward 'blinking.' Both are boys in their teens. Their homecoming, if I may be permitted to use that term, spurred the death of females in and around their presence. Even the most brilliant of doctors are unable to offer a prognosis."

"Wait a minute!" interjected Alex. "You were the only female member in the team and you were the only one to have suffered this acute back pain leading to asthenia. The rest were males and all of them are fit and fine. It was no different the last time too, wasn't it?"

"Yes," accentuated Diane, "and I'm dead sure the insufferable pain occurred only when I looked into their eyes while they blinked."

Rishi listened stupefied to her vociferous vehemence and then answered almost in a whisper, "I think you are right in your postulation. The narrative does defy rationality. It's no wonder then that the Government Health Agency has turned to us for assistance."

Alex intervened, "But there's one thing that flouts my understanding." He paused. "If Diane could perceive the correlation between the ostensibly incongruent indices, 'blinking' and 'back pain' coupled with exhaustion,' then why didn't the other women?"

"Yes, why didn't the others?" echoed Rishi.

"Probably because they've never had an exclusive one to one meeting with them, the way I've had," reasoned Diane. "Probably because it never occurred to them that there could actually be a parallel between the two; probably because ... well..."

Diane wanted to say several things but suffered from a block that stood as an obstacle between her brain and her tongue.

The only thing she said was, "I don't have a convincing answer to that one. But I do know that this calls for a thorough reconnaissance. The doctors, evidently, have perceived a connection between the two, although they are yet to validate it. What else would explain the invention of these unusual pair of glasses especially formulated for their 'blinking' eyes? This child refused to remove them in my presence. He said that his doctor has advised him against doing so."

"Dark glasses? Do you mean to say that even the doctors have found some kind of a connection between the two?" queried Alex.

"I think so," replied Diane. "Why was the boy forced into wearing them, otherwise? Interestingly it is an invention credited to Dr. Memon, the one from Likwi. He, along with another doctor, Dr. Smith collaborated to devise it. I did say, didn't I? That he was simply too brilliant to be practising in such a nameless place."

"That's great to hear," said Rishi. "I admire the doctor. At least it is some solution, temporary though it may be, to lull the potent deaths, but we may be mistaken in thinking that the doctors may have found a tenable connection between the two. They probably have no inkling of what's going on. The eyeglasses may just be a nostrum for the 'blinking' and not necessarily a preventive measure for a conclusive diagnosis resting on an interdependence between the anomalies.

The ebbing of female deaths... the most evident and immediate one at that... is the beneficiary." said Rishi.

"Possibly," expressed Alex. "If such is the case, then it's a saviour in disguise."

"Yes," agreed Rishi. "Whatever, we have to get to the bottom of this. The more I think of it, the more the compulsions that rise in my mind." He turned to Alex, "What's your take on this?"

"There are no two thoughts. It goes far beyond the regularity and interrelatedness of a normal occurrence," Alex reflected ruefully.

"Ask me," voiced Diane. "I've faced the trauma. This is no game but a game changer in the vocabulary of our life. Something somewhere is going the extra mile to bring about the obloquy of mankind. But who? Why? And to what extent? We can't let extrapolated deductions rationalize the conundrum. It calls for 'Action.'"

"And a tangible one, at that. The dreams of the harbourers of this sinister grail have to be crushed," asserted Alex.

The silence that followed was suggestive. Rishi spoke first, "Let's revisit the episodes. You visit these 2 distinctly different places, you meet these 2 'blinking' boys, you suffer from backache and exhaustion, there are females dying all the time in these 2 places and the doctors in both these regions have no clue why it is all happening. Am I right?"

"Those are the facts," insisted Diane with severity.

"Yes, those are the facts," granted Rishi, "facts on which to base our blueprint to resolve the matter expeditiously. We have to conceive ideas that give palpable results - anything with stipulated deliverables. But what? Something's on my mind. Allow me time to think it over. Let's meet tomorrow at the same time to discuss the plausibility of a practical effectuation. Your inputs would be useful to trim the flab out of it. Until then, Diane, I advise you to rest. Your pallor is too pale to indicate that you are in the best of health. And yes, no more of such expeditions for you, in the future."

* * *

Next morning, over a cup of coffee, Rishi pulled out the plan and briefly outlined a selective, edited version of it. He pencilled a schematic diagram for the benefit of Alex and Diane and then proceeded in no uncertain terms, "The Master Plan is mapped to include the all important factor of gender discrimination since we now clearly know with finality that females who interacted with them on multiple occasions died while the males suffered no such mishap."

He looked at the 2 faces staring at them. They seemed interested.

Rishi continued, "Our 'Demo Laboratory' will have to be temporarily modified to consolidate, among other things, the latest technology that will support lifelike dummy models of both genders. They will then have to be linked through connections to enable viewers behind the glass façade of the rostrum, to observe the most microscopic of actions and reactions. By viewers, I mean us. We will then have to lead the boys into the pulpit and allow them to set their eyes on these models. Their reaction will help us decide our next course of action."

"Will it?" responded Alex. "From a layman's point of view I think it's rather puerile."

"Yeah! It does sound audacious," Diane countered. "But we can't be sitting around and twiddling our thumbs forever, despite the doubtful consonance in it. Something needs to be done; no, let me correct myself, something **must** be done." She accentuated the word 'must.' "Females are dying at an inconceivably rapid rate defying the most rational logic and my hunch tells me that these 2 boys are in some way linked to it. Isn't it eerie how they get lost and then suddenly appear from nowhere and the moment they do so, all the females interacting with them, including their mothers, die?"

"You are absolutely right. We can't be sitting here like tethered sheep grazing blindly while a cruel wolf is sneaking up in the bushes," emphasized Rishi.

"Agreed," responded Alex albeit a little reluctantly. "The Plan appeals to me, but I have my qualms about it meeting the objectives. Then there's the question of the boys. Their presence is integral to the experimentation. We will have to get them here. Will they agree?"

"They will have to. They are the protagonists," asserted Rishi.

Diane agreed, "I don't foresee any trouble in Gattu accepting the invitation for a sojourn to the Institute. His Master would understand. Anil? There's a question mark there. His father is far too possessive. I doubt, he'll agree. He is one man who would need a lot of convincing before relenting. It's not a short period, remember."

"He may, if we take him into confidence. We may have to impress upon him the importance of finding a solution for the 'blinking.' Dangling a carrot always helps. Anyway, let's try," concluded Rishi with a firmness that left the other two with little recourse but to give their assent.

CHAPTER IX

It would have appeared odd to the layman as to why the Government Health Agency (GHA) had to lean to 'The Institute of Research for Space and Natural Sciences' for help, but occasionally they did. The Institute was considered the most advanced technologically, highly sophisticated and futuristic. It prided in being the best in the world and it was only prudent that the GHA turned to it whenever there was a crisis. Unquestionably, certain sequels occur naturally in nature, but they are sometimes exacerbated by human activity such as land clearing, urban development, release of untreated sewage, population explosion or decline, and even something as obvious as floods and drought. Satellite data and the associated calculations sometimes helped in resolving such complex issues.

In this case, with very little to go on, a third party opinion was obligatory to explain the number of unusual deaths that were unparallelly big. It was obligatory to seek data from their fleet of Earth observing satellites which tracked human and environmental events that typically preceded such strange phenomena.

At the coffee table Diane, Rishi and Alex mused over the issue. They understood why their inputs were sought and were glad that they had taken it upon themselves to find a workable solution for the inhuman problem that was looming large over the entire humanity. With the macro plans in place, they were now concentrating on the micros to devise a course of action that would help them reach a constructive conclusion.

As Rishi peregrinated through the 'ifs' and 'buts' of his Plan, he realised that 'danger' was the operative word in the whole scheme of things. The extrapolated problems of this seemingly simple nonpareil disturbed him. He looked at Diane distractedly, "Have the boys been brought in?"

Diane inclined her head in assent. "Yes. All 3 of them. They have been quarantined in a separate, discrete locale."

"Three of them?" Rishi expressed surprise. "I thought there were just 2. From where did the third come? Has he been picked up for the same reason?"

Alex too displayed shock.

"Yes," replied Diane with a perfunctory smile.

Rishi gave her an absent-minded look, "I don't recollect him being featured anytime, anywhere in the News. To be honest, I've neither watched the news, or for that matter even read the newspapers for the past 2 – 3 days. Too busy working out the details of the configuration. The scale of female casualties has left me numbed."

Diane understood his state of mind.

"Wake up Rishi! Be prepared to hear the worst," she warned. "There are not just 3, but several; it's just that I've managed to herd 3 of them."

"Several?" Rishi almost jumped off his chair. His voice betrayed his prickling concern. "It's an emergency. What is even more unsettling, is the failure of medical science to come up with a cogent justification for the same. When you first discussed it with me, I merely looked upon it as an eruptive event that would settle soon. Obviously, I was wrong because when the matter came up for discussion the second time, I recognised the audacious magnitude of what was taking place. I was shaken beyond disbelief. And now you say there's a third and many more involved. This is disconcerting."

"Yes. It is," agreed Diane.

"From where did the third one emerge?" asked Rishi. "And what about the others? Shouldn't they be a part of this?"

"We're working on that," stated Diane. "We'll pick them up, one by one. But, right now, it's important to tunnel our concentration on the ones who are with us."

"Where was the third? Which part of the world?" Rishi queried in disbelief.

"Coincidence, sheer coincidence," said Diane. "He comes from a tribal region called Okra, an obscure borough devoid of any distinctiveness to attract media attention, and one which continues to remain unknown and unexplored. Not many have heard of this place."

"Then, how did you?" Rishi said trying to sound nonchalant but only succeeding in betraying that it was an effort.

"You know Steve's friend Ayaz, don't you?"

"Who? That photographer?" This time it was Alex who spoke. "The one who enjoys a brilliant record for turning in unique photographs that consistently have a stamp of the genius, and almost always succeed in miring themselves in controversies?"

"Yes, the same one," replied Diane. "He's a natural; one who needs no script to work on. It happened on one of his rare expeditions while he was toiling for hours to capture a ferocious leopard on the verge of mauling an innocent prey. He succeeded, but by then he had gotten off the beaten track and was lost. Years of working in precarious projects of the kind, had educated him in composure under stress. As single-mindedly as if he were counting the number of trees, he kept marking out the directions which would lead him to his original start point, when he was accosted by a tribesman."

"He does seem to be tangled in weird encounters," intervened Rishi.

"Oh yeah!" agreed Diane. "His life is an encyclopaedia by itself."

"Lucky him!" exclaimed Alex. "Always on an adventure."

"He has his unsafe moments too," reminded Diane.

"That makes me all the more envious of him," said Alex.

"Anyway to get back to what I was saying," proceeded Diane, "Ayaz extended a friendly hand to this tribesman, but his overtures for affability were met with suspicion. Ayaz knew not his language and had to rely on signs and gestures to explain the tight spot he was in. With great disinclination, the clansman proffered to help him find his way out, but not before he had accepted their invitation to spend the night with them. Ayaz consented. When he reached a clearing where the tribesman lived, he was gobsmacked to notice that they were all set to behead a lad of about 16. Cannibalism being rampant in that part of the world, Ayaz's nervousness was understandable. He was unsure if he had been enticed by them as a provision for their next meal."

"And then?" Rishi evinced interest.

"When questioned about their motive for killing, their reply amazed him. He was apprised of an incident when the boy had disappeared for a few days. His reappearance was welcomed with great celebration. He was questioned for hours on his whereabouts during the 'missing' period, but despite their clamorous prodding, they found it implausible that he was unable to recount his disappearance with anything worthwhile."

"That sounds achingly familiar," interrupted Alex.

"I thought the same while he narrated it to me," endorsed Diane. "What followed after that, was not just unbelievable but totally outrageous. Female residents started dying uncannily after complaining of total fatigue. This was sibylline and against all history or experience, since the women of their tribe were particularly strong, far stronger than the men. They failed to understand why all of them started dying without rhyme or reason. The loss was so tangible and measurable that it became a bone of contention for their lineage which they thought was at stake, and the very thought that it would end, compelled them much against their wishes, to abduct women from adjoining locations."

"Are the laws not in place there? Abduction is a crime, you know," declared Rishi.

"I know," claimed Diane. "But this is a godforsaken place, far far from the madding crowd. They make their own laws and contravene them too."

"Men are the same everywhere," commented Alex ruefully.

"So then what happened to the abducted women?" asked Rishi curiously.

"They started dying too," said Diane. "Mistrust set in, primitive instincts predominated, and the boy was under observation. Their qualms were reinforced on observing that all the women who interacted with him died. Beheading the boy was the only way to put an end to these endless and harrowing incidences. It was sheer providence that Ayaz chose to get himself lost in that particular area, and exactly at that moment of time."

"Had I not heard something similar before, I would have dismissed it summarily. But I know better now than to ridicule such seemingly unbelievable anecdotes," stated Rishi insipidly.

Diane smiled, "My thoughts ran on analogous lines. Not Steve though. Over dinner, as Ayaz kept regaling us with his adventures, he listened to him with dissembled seriousness. Steve thought he was spinning a yarn. Not I. I was all ears and then he mentioned something; something that constrained me to attune myself to him with greater mindfulness."

"What was that?" Alex cocked his ears.

"Ayaz sought a reason, from them, for putting the total onus on the boy. When the boy reappeared, they said, they were enraptured by the strange way in which he blinked 3 times in succession in a mechanical fashion. Constant monitoring led them to materially strengthen their conclusion. It was difficult to overlook the fact that every time a woman looked at him she would cry out in pain and say she was tired. A dozen similar interactions later, the relevant lady invariably died. The males remained unaffected."

"Unbelievable!" Rishi exclaimed. "I wonder at the boutade... it seems to have assumed dangerously disproportionate levels with its tentacles spread all over the globe."

"Yes, it's a denunciatory ball game," conceded Alex.

"What happened next?" prompted Rishi.

"Ayaz offered to take the young boy away. They relented. Ever since, the boy's been residing with him. I advised him against letting the boy intermingle with anybody, especially the females, prudently adding credence to my advice by giving him an account of

Gattu and Anil. It was then that both he and Steve remarked that they had been reading about it. Ayaz couldn't contain his excitement. And Steve... Well! The less said about him the better."

"Yes! He's one helluva guy. Very hard to please," opined Rishi.

"I know. To cut the long story short, I requested Ayaz to fetch him to our Institute. My gender put me in no mood to get him here myself, certainly not after those 2 or 3 interactions which have left me virtually incapacitated with this slight back pain and exhaustion. Since the plans to conduct our experiment were already afoot, I thought it would be a seemingly good idea to make him a part of our trial," concluded Diane.

"Good job!" commended Rishi. "Has care been taken to counsel the boys into wearing that peculiar eyewear? You did mention that it has proven to be instrumental in inhibiting adverse reaction. What's that boy's name? Ah yes! Anil! He has been sporting something similar for some time now, isn't it?"

"Yes. And so has Gattu, although I got to know of that much later. In fact, I've been told that Gattu was the first. Remember Dr. Memon? I think I mentioned earlier that this invention has been credited to him and his colleague Dr. Smith. Gattu belongs to his place. We have given a new pair to each of the boys to avoid unfortunate repercussions. It succeeds in minimising the adverse effect, but doesn't hinder a vindictive reaction totally. It has to be replaced every 2 weeks."

Rishi acknowledged with a slight nod of his head, and Alex with an inaudible grunt.

A sharp silence trod on the heels of this exchange. Diane punctured it.

"When are we starting?" she asked.

"In the next 2 days. The laboratory is being assembled. Two robots - one male and the other female are a part of the configuration. They have been structured to appear as lifelike as possible."

"Excellent!" Both Diane and Alex exclaimed unanimously.

* * *

The chassis made of glass was fitted with world class facilities. It included new measurement concepts, technologies and system coded capabilities to ensure the visualised success. Six different science instruments were employed to collect the uttermost miniscule data and measure discrepancy of the most infinitesimal magnitude. The 'Observers' section enabled easy visibility to those seated there. The entire provisional set-up was pieced together to make it conducive for the dummy run. It was unlike anything before, and would hopefully give a vital breakthrough that was being sought.

"Great!" said Diane when she set her eyes on the temporary technical edifice. "Brings to my mind the time when people were contracting a strange disease for which presumably there was no explanation. What was it called?"

"The 'Martian Measles,'" Rishi replied.

"Yes," affirmed Diane. "The GHA had turned to us for help. But I found the name a little odd."

"It was so nicknamed because of its alien symptoms," explained Rishi. "They had sought data from our fleet of Earth observation satellites to track human and environmental events that may have preceded its outbreak. But it was different then. There was no such elaborate set-up. The demand, at that time, was merely for digital property maps to help them in their research. They needed no other assistance from us. Here our responsibility extends beyond that. We have to find the source responsible for the anomalousness."

Alex entered just then. "O Wow!" he exclaimed looking appreciatively at the mechanically compiled electronic humanlike figures that had been uprighted on a huge stage, a few metres higher than the ground level. "Are those for real?"

"An admirable job isn't it?" said Rishi. "They look almost human, but they are not the only ones who are a part of this. We have kept in abeyance 2 real human models. We intend sending them alternatively to gauge the difference in reaction between inanimate and animate objects, when subjected to the scrutiny of the boys."

"Impressive!" replied Alex. "A remarkable job by Shelar and his team. But how did you manage to convince the live ones to be a part of this, especially the female?"

"I didn't. They volunteered. In fact, the female was more enthusiastic. Both the male and the female are a part of Shelar's team. I did explain the repercussions to the lady, but she was defiant. She said if Diane was willing to risk her neck, then why not she. Besides, she was aware that females died only after multiple reactions, and not just one. It pays to upgrade one's knowledge."

"A brave one, I must admit. Hats off to her," said Alex in a deferential tone.

"Yes. Diane's her role model."

"Is she?" stressed Alex incredulously. "That's a heartening piece of information. There was a time when young women looked up to entertainers like movie stars as their role models. Now they look up to intellectuals. Wow! That's a great change - a change for the better. Anway, when do we begin?"

"Any instant."

It was the first time in the history of the Institute that they were working on a practical substitute instead of formulating abstract convictions based on tiresome formulae that were beyond the comprehension of the common populace. It was like returning to the primitives. The moment was convention defying, and the underlying tension palpable.

While Shelar and his assistants got busy with the placement of the models, Alex, Diane, Rishi and a few others seated themselves, comfortably, in the mini glass auditorium directly in line of vision of what

Rishi jestingly referred to as the stage. Last minute instructions dwarfed all else. "There are 2 things I wish to emphasize. One, the distance between the boys and the models should not be more than 15 feet... 2, the line of vision ... the boys should be able to look directly into the eyes of the models; that's of supreme import," mandated Rishi.

A concerted nodding of the heads indicated that his instructions were well taken.

The cameras whirred and the automated clicking of instructions onto a keyboard could be heard. A buzzer sounded, and a red light blinked as Shelar started counting backwards... Three... Two... One... Go... Instantaneously, the boys entered the acculturated glass environment and gaped in awe at the mute models who uncannily resembled ordinary mortals, but were bereft of expression and contouring so typical of humans. Their movements were closely monitored and their images simultaneously captured, to be studied later, for error correction. The crucial part of the Plan was to elicit response from the archetypes. But, that didn't happen. They did not react as expected. Rishi was sorely dissatisfied. His baby had failed to deliver results.

"No reaction," commented Alex rather sceptically. "I hope this has not been a waste of time. Are you sure we have not adjudged incorrectly?"

Diane shared his feelings.

"There should have been some reaction," Rishi replied. "I was thorough with the details. Their curious 'blinking' continues but there is no response from the robots. What could have gone wrong?" He was visibly flustered.

"Despite the well-thought-out clichés, a project of this kind can easily run into bad weather. There's a possibility that something may have been unconsciously overlooked. Should we rerun the demonstration for any micro detail that may have been missed?" prompted Alex.

"No. Not until we have concluded it in its entirety," said Rishi. "This is the first phase. I may have been misled into thinking that an outcome would be expected at this juncture itself. Let's run it further and judge success only after the conclusion of the second phase."

He signalled Shelar to move on to the next instruction block which entailed taking the prototypes out and letting the 'live' ones enter. "Please be advised that the live models are made to stand exactly the way the dummies did," he announced.

"Right, Sir."

The 'live' models made a dramatic entry and strategically posed in a manner expected of them. The disenchantment that had set in during the first part of the experiment did nothing to dampen the enthusiasm of the small but distinguished audience. They waited with bated breath for the second part. And then... the female reacted. She ostensibly swerved forward a little while simultaneously letting out a small cry of pain.

Diane sat upright. "Quick," she responded, "turn off the lights and allow darkness to descend. Be swift and lead her out of the 'stage' before things take a turn for the worse. She's flesh and blood."

The chamber which until then had been injected with prolepsis was soon shrouded in tenebrosity. The boys were led back to their quarters. The female was escorted and entrusted in the safe hands of the resident doctor. "It was awful," she recounted as the doctor proceeded to test her pulse and heartbeat. "Indescribable to say the least! I feel exhausted and breathless. And that pain..."

Meanwhile, the temporarily assembled laboratory pulsated with excitement. Rishi was awfully pleased with himself. So enslaved had he become to all the computer paraphernalia that he had gradually begun to doubt his cerebral ability. His intellectual self-esteem was gratified to note that his brain could, yet, be a conduit for simple ideas that could give tangible results. His reverie was intruded upon by Diane who could barely contain her elation. "This was exactly how I reacted when I saw the boys, the first time. What we beheld now only confirms what I had maintained earlier. These boys are indeed responsible for the deaths. There's something appalling about the fact that it's only the female who is being attacked. Did you notice how the male remained unaffected?"

"You are right," agreed Rishi. "It is indeed eccentric. Not just eccentric but menacing."

Rishi's gratification on the success of his experiment was marred by the thought that he could not apprise the others of a rational reason for this gender discrimination. Besides, he was concerned about the fate of the human female who had become a victim of his niggardly exploit. He was in no mental frame of mind to think coherently.

"We can defer our fantasizing for another time," Rishi told Diane. "I'm in no good mood now, not after what happened to that brave young lady. I hold myself wholly responsible for her affliction. But was I left with an option? No. The dummies failed to respond, and it was pertinent to see if the live ones did. It would have been injudicious of me to obviate her entry. And you observed, didn't you? The living individual reacted, where the robots failed, proving that it was an absolute inevitability."

"Rest assured," said Diane. "The twinge, as I prefer to call it, will not necessarily fade away but it'll weaken and I'm talking from my individual experience. Besides, as you rightly stated, she, as a living individual, reacted while the robots didn't. Her entry was crucial for leveraging our inference. Do you recall my narration after my return from 'Likwi'? I had mentioned a similar throbbing when I set my eyes on Gattu for the first time."

Alex frowned and then beamed, "You mean she'll be fine?"

"Not exclusively but she would be able to carry out her daily tasks comfortably, without any impairment.

But in our best interests, it is crucial to avoid subjecting her to any further scrutiny of this kind. The after-effect is bound to be more severe, the second time. And we all know the unsaid when several interactions come to pass. Let her rest awhile under the doctor's supervision. We can arrange to send her back the instant she feels revitalized."

"Let's pay her a visit," suggested Alex.

* * *

Alex explained. Others listened. They were seated in Rishi's cabin. "Her cry of pain was juxtaposed by a slight bend of her back. The actions were simultaneous," mused Alex. His outline was synoptic and epigrammatic.

"I encountered the same during my meeting with both Gattu and Anil," confirmed Diane. "In fact, at that instant, I thought I would never be able to walk straight for a long time. But I did; thanks to Steve and his meditation techniques."

"Yes, you did mention something on that subject the last time you returned from Likwi," said Rishi. "But we didn't have much time to elaborate on it. We were too engrossed in discussing our trip to Hosuru. Now that we are at it, maybe you can tell us a little about his sudden interest in meditation. Where did Steve pick it from? Some new destination?"

"You guessed it right," said Diane. "He's always been like that. Trying to pick up something new every time he visits an unexplored destination; he refers to it as 'inspiration.'"

"So, this time it is meditation, is it? For how long?" questioned Rishi.

"Not sure," replied Diane. "Generally, such *'AHA'* moments are short-lived, but not this time; looks like this has come to stay."

"Everything else has been swept aside, is it?" asked Rishi in disbelief. He knew Steve. He was the kind who got bored easily with everything but his humanitarian causes.

"No, no!" stressed Diane. "His heart continues to be with his social causes, but he has hooked on to this thing called 'meditation' in a big way. In fact, it has now become a part of his daily ritual. His claims that it has helped him profoundly in retaining vigour and health, can easily be authenticated by his present physical appearance."

"Really?" asked a surprised Rishi. "The last time I met him he complained of not getting adequate time to pamper his corporal self. He blamed it on his busy diary."

"That was almost 3 months ago," replied Diane. "You should see him now. Even the younger guys envy him. According to him, meditation is a science. If practised regularly with the right technique, the health benefits are tremendous. Our body benefits both mentally and physically, or so he says. I am beginning to realise that he is right. His food habits have changed, and he has begun to take his exercise regime seriously. As for me, I do get quite a bit of relief after a few minutes of it. It's a simple act that can be done by anybody at any time."

"That's heartening," responded Rishi. "We must discuss this later. Not now. It has to take a backseat to the mystery of the 'blinking boys.'"

"Yes," agreed Alex who all along was listening to the dialogue with rapt attention. "We are veering from the topic, but discussions on stray subjects like this, help relieve the oppressiveness of the situation."

Rishi endorsed his statement. "Yes, they lighten an otherwise heavy-laden atmosphere."

Rishi then looked at both Alex and Diane with a question writ large on his face.

"Want to say something?" asked Diane.

"Yes. You guessed right," affirmed Rishis. "There's one crucial point that keeps pirouetting, in my mind."

His expression pleaded unsuccessfulness.

Alex and Diane gave him a sharp look.

"Did you notice?" Rishi said laying his thoughts across. "The real human beings reacted. The steel robots didn't. Why did they not? They had been crafted to perfection without missing out a single detail - even to the extent that the heart pumped at 72 beats per minute, like all normal human beings do."

"There's a stark difference between metal and pure flesh, blood and bone, Rishi," reminded Diane gently.

"I had conceded that while assembling the robots," justified Rishi. "But that's not what puts me in a pickle."

"Then what does?" prompted Alex.

"The thought that the hand behind this sinistrous game plan knows the human body comprehensively... so comprehensively that it knows the subtlest of difference between the two genders, and the minutest of contrast between the human flesh and other metals. Isn't that a frightening thought?"

Grim faces indicated that the point was well taken.

"It's graver than we thought," proceeded Rishi. "We must ask ourselves questions, questions and even more questions. We must do more, and even more in depth research. Only then, perhaps, we may reach a logical conclusion. I sense subterfuge; I sense the tightening of the finger of a ruthless killer on the trigger of a gun. If the finger succeeds in hitting the bull's eye, we are looking into a future of no return."

The thought was unnerving. It made their hair stand on end.

"Do you think it would help if we stage the experiment with the robots again?" suggested Diane. "It's always advisable to revisit. There's a likelihood that our annotations the second time will reveal something that has not been noticed in the first round."

"That would not be necessary. Some weighty brainstorming is all that we need to." said Rishi. "The images have been stored in the magnetic spool. We'll run and rerun them. There's every possibility that a seemingly indistinct detail may have skipped our scrutiny."

"Good idea!" said Diane. "When do you think we can begin?"

"In the morning. Tomorrow. It's the weekend and nobody will be around. Just you, Alex and I. I can request Shelar to send his team for a while," said Rishi.

"And Ayaz too," offered Diane.

"Who?" Rishi asked.

"Ayaz!" repeated Diane.

"The same photographer who brought the third guy from some remote tribal area?" reaffirmed Rishi.

"Yes."

"How would his presence help us?" intervened Alex. "Besides, a third party incursion is not appealing."

"Yes, secrecy is the need of the hour," subscribed Rishi.

"I know, I know. You are justified in putting your foot down and disagreeing with me," said Diane rather weakly. "And I, in no way, take offence for brushing aside my request. It's just that he is technically far superior to us in this field. His accomplishments in several cases remain unrivalled and nobody knows the ropes as well as he does. Many even refer to him as a photographic genius despite his idiosyncrasies and revolutionary theories. I thought, perhaps, his point of view may direct us on an alternate track altogether."

"Are you sure?" asked Rishi.

"Certainly yes," impressed Diane. "Don't forget he's a photographer par excellence, and photographers are blessed with a predisposition to view things differently from the rest of us."

"Well! That does make sense," admitted Rishi, a little reluctantly, "but can we trust him? It would do us no good if the world knows of what is transpiring within the confines of our Institute. Rumours, fake news, false propaganda would only assist in diluting the quality of our work and creating unnecessary obstacles and delay."

"Of course, yes! He can be trusted, absolutely. I can vouch for that," Diane replied, a little offended by what Rishi stated.

"Would he be in town, though?" enquired Rishi. "Of the little that I've heard of him, from you, he's always travelling."

"Fortunately, he's around," confirmed Diane. "Besides, I'm sure he'd be excited to be a part of this; more so, because one of the boys is his protégé."

"If you think he is trustworthy and his presence would help us, then I have no objections," Rishi replied, rather blandly. "Tomorrow then; around 10.30 in the morning."

* * *

Ayaz accepted Diane's request with a zeal beyond her expectations.

"You mean to say, you need my presence. Wow! Interesting! Sure, I'll be there. Thanks a lot Diane," he said giving his consent in the garrulous fashion that was so typical of him. He had often wondered if his naturalistic insights had brought about great changes to the way things existed in the world, or if they were merely restricted to coffee table conversations. Maybe this time around he would know.

As Ayaz moved past the formidable security at the Research Institute, he was dumbfounded at the idyllic, picturesque setting. He recalled the figures of speech his English teacher had taught him, while he was in school. That was a long time ago, but he still remembered them. They always drifted in his subconscious during all his photographic sessions. His favourite then and even now was 'imageries'... What were they? Ah yes! Tactile, visual, olfactory and gustatory. This place aroused all his senses.

His conversations with Diane had led him to presume that the place where she worked was drab, uninspiring and prosaic. But what lay before him defied all imagination. The foliage was not just profuse and deific, but a paradise fit for the Gods. The architect who had designed the landscape apparently had a great eye for detail and the caretaker or gardener who maintained it was doing a marvellous job of retaining its breathtaking beauty. He wondered if the people who worked here even realised how fortunate they were.

"Shall we go in?" asked Diane awakening him from his trance.

"You didn't tell me you worked in such an exquisite place. It's heavenly," said Ayaz.

Diane looked around. So accustomed was she to the place that it never ever occurred to her that the surroundings were worthy of admiration, not until Ayaz mentioned it. "Come on, Ayaz," she retorted with cynicism, "you must be joking. Do I detect a tinge of sarcasm from a man who has travelled around the globe and yet chooses to make a mention of this place as being out of the world?"

"View it through my eyes to appreciate the panegyric," replied Ayaz enthusiastically.

Diane spanned her eyes to cover the stretch of land in her vision. Ayaz was right. She did indeed work in one of the most beautiful environs and wondered why she had never given a thought to it. She blamed it on her work.

"If ever there is an article to be written on the Institute based on its aesthetics, chalk me out as the photographer. I bet the article will become the most read, the most popular," Ayaz stated.

"All right, all right," Diane said a little impatiently." Come, let's go in. Everybody's waiting for you."

The preliminary introductions being taken care of, Rishi recited his instructions to Shelar, "Play out the images at a normal pace initially, then a little slower, then even slower than before and finally freeze every frame so that we can dwell on every trice that has been captured. I hope I am clear?" he said. He then signalled for the lights to be switched off.

The first screening was a damp squib. It ended with no comments as there was nothing worth a mention. The second screening was a repeat of the first. Nothing special. The third screening was at an even slower pace than the first 2, and when they were merely 95 seconds into it Ayaz stiffened and gave Diane, a sidelong glance, "Notice something?"

"Not really!" she said.

The photographer in Ayaz dominated. "I think we should rewind and replay this section again at an even more unhurried pace. Stop when I ask you to, and then freeze the frame," he told Shelar.

The projector whirred and the images moved smoothly on the display, until the hand that held the contraption steadied to a halt when Ayaz commanded. He looked excitedly at Diane, "See, can you see what I see?"

Diane sounded unsure, "Yes, I think I do but it is barely visible."

"Anything exciting?" intervened Rishi.

"Just immobilize the shot that Ayaz is referring to. Let's see if he and I really did see what we think we saw," Diane replied.

"Okay."

Everybody gazed at the relevant snapshot with their mouths agape when it was frozen. A faint red streak appeared to dash from the boys to the models; so faint was it, that one would miss it if one was not acutely vigilant.

"Let's play that backwards in slow motion to trace its origin," demanded Rishi.

"There it is," screamed Diane almost falling off her seat, "it appears to originate from the eyes."

"Exactly. And the blinking happens simultaneously. It then reaches and ends at the models' eyes," added Ayaz. "There is no reaction from the robots."

"Let's retrace the film where the real models were employed, at a similar pace," suggested Rishi.

A long impregnable hush accented the unbelievability of the viewers, as they watched a replay with the live models. The moment was earth-shattering.

"This is jaw-dropping," Ayaz exclaimed. "I can't believe it's true, but it is. I've actually been able to mark the pathway of this mysterious red streak. It starts from the eyes of the boys hurtles through nothingness and reaches the eyes of the female human model who instantaneously bends slightly and gives out a barely audible shriek; all in a matter of a fraction of a second. It's lightning quick and extremely obscure. One has to strain oneself to notice it. Strangely, the same red streak has no effect on the male human model, although it does reach his eyes too."

"Does it happen with all the boys?" asked Rishi who appeared to have noticed nothing. Perhaps his eyes were not as sharp as the rest. Even Alex appeared to have noticed it.

"Yes, it does," replied Ayaz. "And another thing that's noteworthy is the colour of red... Prima facie it is amazingly different... almost alien. It's outre. If you could arrange to give me hard copies of the photographic scripts, I would be better able to scrutinise them assiduously. In the meanwhile, I suggest that we put these boys through a scanner and see if the X-ray prints reveal anything."

***

The scanned images that reached Rishi's desk the following morning revealed nothing unusual except for a deviant bright patch around the region of the right ear. It was encircled to draw his attention. He perused the accompanying report for comments:

*The bones of the skull are normal in size and appearance*No foreign objects or abnormalities present*No broken bones present*No sign of any disease, tumour or endocrine disorders*No sign of birth defects.*

At the bottom was a note in bold letters with an asterisk to highlight it **'A freakish lucent patch above the right ear which has not been noticed to date and is unique to all the 3 images could be a cause for concern. CT scan recommended.'**

He called for the radiologic technologist, "Is this the best you could do? And why have you recommended a CT scan? Your report says that there's nothing abnormal except for that patch."

"You're right," responded the technician. "Superficially there's nothing uncommon about them, but while scanning I sensed some interference, the kind which generally happens when the rays encounter a metal object, but as you can see, except for a bright patch, there is no metal object there. It's something I have never observed before."

"Really?" said a surprised Rishi.

"Yes," affirmed the technician. "The detector frictioned an instant there, and then moved. Something that never happens. It was quick - as quick as lightning to use a simile. There, and yet not there. If one lacks the kind of experience I have, one would barely have sensed that. I've deliberately circled that portion to guide you."

"Does it occur in all the cases?"

"Yes, it does."

"Where exactly did you sense this intercession?"

"The same place that I have marked; around the region of the ears."

"A remarkable observation!" exclaimed Rishi. "Ayaz also mentioned that the red streak originated from somewhere around there. Well! Not exactly. He pointed to the region of the socket that was very close to the ears. Just one more request. Can you arrange for a CT scan and send the reports to my office tomorrow morning?"

"I'll do that," complied the technician.

"Thank you."

The morning that followed was bright but charged with questions that sought evasive answers. Both Alex and Diane were sitting in Rishi's cabin when he entered. "Anything?" they vocalized together.

"Maybe, maybe not," replied Rishi.

"What do you mean?" asked Diane.

"Let him get the images. I shall explain."

When the images were brought in, Rishi pointed out to a small patch slightly above the right ear. "Notice an ultrabright smudge here? The technician explained that dense tissues in the body like the bones block X-rays, and look white in the picture and so do metal objects, if any. He found something contradictory. The speck that he has encircled for our benefit, appears bright, indicating the presence of a foreign metal object. But strangely, it does not show itself in any way. He also educated me to the fact that radiolucent objects don't show themselves in images. Metal and radiolucency don't gel. He appeared confused. He has never seen something like this in the past. It has been saved in the data base for further study."

Alex and Diane peered closely. Diane was the first to react, "It's hardly distinguishable."

"You're right," Rishi replied. "It's hardly noticeable, but it is present in all the 3 images and is peculiar to them only. Never has something similar been observed in the past."

"So?" asked Alex.

"Think back to what Ayaz had said during the viewing. He spoke of a hazy reddish streak starting from somewhere around the region of the eyes of all the 3 boys. There is an origin and there is an end and then there is a transmission from the origin to the end."

Alex and Diane nodded their heads in unison.

"Isn't that food for thought?" questioned Rishi.

"Guess, you are right," said Alex. "There must be some correlation. That is the patch that we need to concentrate on."

"How do we do that?" questioned Diane.

"These were the best that the technologist could do for us," replied Rishi.

"Would surgery be an option?" queried Alex.

"What surgery?" Rishi asked.

"Since nothing further can be revealed by this scanning gizmo, do you think we could request a surgeon to work on that portion?" explained Alex.

"Joking, aren't you? Is surgery a child's play? If these copies reveal the presence of nothing; just a blank space, there's not a remotest chance that a surgeon will pamper our wishes," asserted Diane.

"A surgeon would be better poised to give us an opinion on that; someone unconventional; someone willing to take risks and someone as enthusiastic as you and I," said Alex.

"Even if he does give a nod, a casualty at this juncture is not a pretty thought," replied Diane.

"I agree that a fatality at this juncture is not a pretty thought but on second thoughts, it's not a bad suggestion. Worries need to be brushed aside if there exists even the remotest possibility of achieving a breakthrough," assured Rishi. "Precautions will have to be taken, and assuredly they will be."

CHAPTER X

Mr. Varma looked distraught. The Inspector felt sorry for the gentleman sitting in front of him. It had only been a few days since he had lodged a complaint about his son Karan, and here he was filing yet another one for his second son Kunal who had apparently vanished into nothingness just the way Karan had. His mind fleetingly recounted the events that led to this moment. Mr. Varma had 2 sons. They were twins and all of 16. His wife and he were relaxing in their bedroom on a Saturday evening when Kunal, the younger son, came running to tell him that Karan was nowhere to be seen. Mr. Varma looked at Kunal searching for traces of mischief in his glinting eyes. The boys were notorious in the neighbourhood for their pranks. "Are you sure?" he asked more in affirmation rather than in interrogation.

"Yes papa," replied the boy. "We were building a bird house in the garden outside when we realised that it had become quite dark. I ran into the house to switch on the garden lights. When I came out, Karan was not there. I ignored him. You know how he is, no? Always bullying me. I thought he was up to his usual tricks. I carried on without his help thinking he would show up any moment and start harrying me by throwing tantrums and breaking the parts of the house that I had so systematically built. But he didn't. I looked for him at the usual hideouts. He was not there. The front gate down the path leading towards the road is latched and closed. That means, he has not gone out either. That is why I came to tell you."

Mr. Varma was shell-shocked.

The bungalow where the Varmas stayed was one of a series of red-tile-roofed houses that lined the entire stretch of a street that led towards the highway. These homes were sited in the suburbs of a fashionable metropolitan city and housed several Government officials. Each abode consisted of a ground and an upper storey. The master bedroom was in the upper storey. Mrs. Varma was all flustered. She and her husband ran frantically downstairs and scouted feverishly around the garden - a small little place with quaint flowers that Mrs. Varma had painstakingly nurtured to give her bungalow a pretty look. It took them barely 10 minutes to conclude that Karan was not there. In a daze, Mrs. Varma walked towards the compound wall and called out to her closest neighbour Mrs. Singh.

A gregarious lady by disposition, Mrs. Singh was as generous as she was suave. Her kitty parties were the only events that disturbed the otherwise lazy existence of this sleepy place that was characterised by warm hearths, well fed pet dogs that hardly barked, women who knitted in the late afternoons, and sporadic ritual lunches in one another's houses. They were great opportunities for the ladies to mollycoddle themselves in some frivolous gossip, while at the same time, indulge in tasting some of the rarest global cuisines. Despite the dominance of modern conveniences and literary achievements, the lifestyle here, was out of sync with the bustling metros. Few things rustled the somnolence of this unique but charming mise en scene. So, when she heard the anguish in Mrs. Varma's voice she scampered out with great speed.

"Hello! Is anything wrong?" she said looking at Mrs. Varma.

"It's Karan," replied Mrs. Varma. "He is not to be seen. Has he come to your house, by any chance?"

"No," replied Mrs. Singh. "Have you checked the gate?"

"Yes, it's locked."

"Then how could he be here?" asked a puzzled Mrs. Singh.

"One never knows with him," stated a harried Mrs. Varma. "He may have, as usual, jumped over the compound wall and sneaked in. He often does that, you know."

"I know, but he's not done that today, at least. Have you checked with the Chopras?"

"No. I'll do that, right away."

Mrs. Chopra replied in the negative when asked a similar question.

"Then where could he have gone?" Mrs. Varma moaned in a voice that expressed her agony.

In a matter of a few seconds, the entire neighbourhood was awake. A frantic hunt was launched for the boy who was apparently lost. Consolations streamed in from all quarters and people offered to help in any which way that they could. Opinions were bandied, suggestions were accepted and parried and yet, at the end of all the commotion, the situation continued to lack clarity. Finally, it was decided that filing an FIR, at the local constabulary, would be the most apposite thing to do. Mr. Varma, accompanied by Mr. Singh and Mr. Chopra, walked the short distance to the police station.

The Inspector, an elderly gentleman with a mop of thick dark hair, enjoyed immense popularity with the residents as he was ever willing to lend a patient ear to anyone in trouble. He expressed genuine concern when acquainted of the problem and, along with his trusted team, set cracking on the complaint immediately. There was no trace of Karan. Mr. Varma had seen and heard on television of young boys disappearing without trace and returning, a few days later, with a strange case of 'blinking eyes.' He hoped this was not the case with his son. He expressed his presentiment to the Inspector who reassured him.

"Oh, no, no," the Inspector bellowed, "I am sure Karan will be found soon enough. In the meanwhile, I suggest you go home and rest awhile. You need it."

Karan was yet to be found, and here was Mr. Varma lodging a complaint for his second son Kunal who, like his brother, had gone missing too.

"I am perplexed," said Mr. Varma. "I kept watch while Kunal played in the garden. I hold myself responsible because I had forced him to go out and play. He was hesitant. He had become insecure and fearful ever since Karan went missing. It took a lot of coaxing on my part to encourage him to go out while I kept watch through the window. It was but for a few minutes that I had hesitantly gone to fetch my pipe while concurrently warning him against going towards the gate. I returned, and he was gone."

Mr. Varma paused to sip some water offered by the Inspector in a show of solidarity.

"Sounds far-fetched, doesn't it?" continued Mr. Varma with his outpouring. "And let me assure you, there's nothing wrong with my eyesight for sure. My wife has gone insane and her sister, who is visiting us on the pretext of taking care of her, is no better either. I don't know how to console them. When Karan went missing, somewhere, just somewhere, I harboured a hope that like the other boys who had gone missing my son Karan too would be found, but I presume that's not the case here. On the other hand, my second son goes missing too. Nothing can get worse than this. All is lost. I don't know how to handle this. I have no will to live any more. What do I do Inspector?"

The Inspector looked down, trying to hide the tears welling up his eyes. He was unsure of what was expected of him in terms of consolation. A slight bow of his head was the only response he could muster.

When Karan had gone missing, he had left no stone unturned to find him. The police stations, in and around the vicinity, were all notified of the incident to bring the matter to a satisfactory closure. The Inspector had even gone to the extent of enlisting help from an old acquaintance who worked in the President's office, but had drawn a blank. And now he was faced with yet another dilemma; that of filing an FIR for the young lad's brother. He was at odds. He looked around here and there merely as a perfunctory gesture for want of something better to do. Two were now lost. Two from the same family.

Arrangements were made to broadcast the news and put up the pictures of the 2 boys, but there was no useful information forthcoming from any quarter.

The immediate area became alert and children were coerced to stay within the 4 walls of their houses. No one knew who would get lost next.

*** * ***

"Care for a cup of coffee, anyone?'" asked Rishi, as he lifted his head and looked at Alex and Diane. The nods were in the affirmative. The trio's initial joie de vivre had made way for despondency. The shocking disappearances and reappearances with abnormal eyes had led them to a different world altogether. Their interest had been heightened to a danger that was fathomless, and for a brief while they had been confident that they would swerve it off their track. Not anymore. The surgeon had ruled out the possibility of any alien element being embedded in the body despite the technician's observation of a controversial abnormality in the scan. His opinion had dampened their enthusiasm.

"I'm aware it's a crisis calling for pressing attendance to the exclusion of everything else, but honestly, I do not know from where to start," said Rishi, feverishly. "Every time I divine that we are on the right track, a fresh development intervenes to take us back to square one."

"I confess I have no idea either," replied Diane. "Like the spider that never fails to spin its web, every time it is disentangled, we will have to start all over - not once, not twice but over and over again. There's a lot to learn from these tarantulas. They weigh nothing at all and yet tire not to respin their web every time, and all the time. Our case is no different; it is as intricate and perplexing. We will have to accommodate and reaccommodate ourselves every hour, every day to find an appurtenant solution."

Alex waited for the enormity of Diane's words to sink in. The task was indeed Herculean. Through the opprobrium of dubiety and exasperation he was aware of Rishi appraising the situation mentally and feeling as powerless as he. Of what use was knowledge if it could not be utilised at the opportune moment, of what use was training if it could not be applied in apt situations, and of what use was the so-called human intelligence when it could not be commanded under trying frameworks.

Alex felt incompetent, ineffectual and inept. Perhaps, they needed to travel on an altered path altogether. They generally based their rationale on hard facts and analysis, but this was one instance where he thought, they would have to tilt their reasoning more to an intuitive process rather than a diagnostic one because they were seldom confronted by incomprehension of this kind. After much introspection all he said was, "Has Ayaz given his feedback?"

"Not yet," replied Rishi.

"And Sridhar? Have you conferred with him?" Diane asked of Rishi.

"Yes, I did. He mentioned very briefly that he had observed something unusual in recent times but had brusquely brushed it aside in the light of the vagaries of the atmosphere and its dangerously high pollution levels. It was only when I stated that we were at a dead end and needed some fresh perspective to reach a concrete conclusion that he agreed to dwell on these glosses more seriously and discuss it with us, but only after he had personally reconfirmed his observations. He didn't want it to be a wild goose chase," replied Rishi.

"And when would that be?" asked Diane.

"Frankly speaking, I don't know and he doesn't know either," replied Rishi.

"He doesn't know?" asserted Diane with a crestfallen look. "He should have given us some deadline."

"It's not as simple as you think. The case is so complex, that he would, most definitely, need time to reach a level-headed inference," indicated Rishi.

The taciturnity that engulfed the room after this brief conversation was long and pregnant. The portentous ticking of the clock on the wall only added to the ominousness of the state of affairs. After what seemed like eternity, Alex spoke. He summed up the situation as briefly as possible and then, after yet another long pause, called for an adjournment. "It's best to leave with the persisting optimism that the brain will be a fertile bed for avant-garde ideas tomorrow," he concluded.

"Perhaps you are right. Let's sign off for the day. It's past 8," said Diane as she studied her watch.

They walked with heavy hearts towards the parking slot. The gentle cool breeze that flirted shamelessly with the well maintained greenery of the campus was a soothing balm to their tired, saturated minds.

"Very pleasant," remarked Diane frivolously, merely to lighten the mood of the moment. "Ayaz was right in saying that we worked in paradise. I wish our workstation was out here. The idyllic atmosphere would have recompensed for the chaotic turmoil in our brains."

"Night time, Diane. It's night time, and that is why the weather is so pleasant," reminded Alex. "The afternoons are a deep contrast. It is so blazing hot then, that working here, at that time, would be a sweltering affair. Try once, and you will exude gratitude for the cool air-conditioned clime that you work in."

"I know, I was joking," jested Diane.

"By the way, what's the latest take on the subject?" butted Rishi who was still in an introspective mood. "My involvement with this project has distanced me from everything. I know it sounds unbelievable, but it's true that I've not even read the newspapers for the past entire week."

"Boy! Am I glad you brought up the subject!" said Diane in a voice that was suddenly filled with excitement. "How could I have forgotten? The news is truly unreal. I had planned to tell you about it, first thing in the morning, but it slipped from my mind. Mass media is on an overdrive to give it maximum coverage. It's ironical that we who are involved in it in the true sense are being distanced from it. I mean, nobody has contacted us for our view."

"If it's yet another anecdote of a boy being lost and found again with blinking eyes, why would they?" said Alex sarcastically. "They would know exactly what we would say."

"It's different this time," said Diane. "It's not one but two boys who are lost now; twins. The elder went missing first, only to be followed by the younger a few days later. Both, presumably, were playing in the tiny park attached to their stand-alone house. The front gate was locked. The local police presume that it could be a case of abduction, although there is no concrete circumstantial evidence to confirm the presence of any stranger who may have been responsible for their kidnapping, at the time they were assumed to be missing."

She paused for reflection and then proceeded, "Come to think of it! Maybe that's the reason why they didn't bother to get in touch with us. They think it's a case of holding these lads hostage for ransom. Their father is a prominent government official. The news is hogging special attention on the national coverage. The 'National Detective Agency' has been forced to step in and proclaim a special interest in the case. I personally disagree. I think it is the same stuff with an intriguing twist to it."

"My God!" Rishi restrained a slight sigh which was a vexatious murmur barely above the threshold of audibility. "Now the number has risen to 2, is it? How dramatic! This is getting to be more and more stymieing. I shan't be surprised if 3 months down the line, we will be discussing the disappearance of 3 individuals... then 4... then 5... the drift is daunting. Like the first trickling drops of rain... seemingly innocent at first but suddenly gushing into a huge flood. Our suppositions and exertions will boil down to nothing, if the hand behind the mystifying evaporations is faster and errorless, despite a high degree of heedfulness. Yet again we'll have to throw the towel over our existing theory and start from the roots all over again. If this persists, we are doomed. I feel worthless."

"Worthlessness is too weak a word," intervened Alex. "The addition of a new feature to the poser with every passing day, is only adding to the existing mess. I find myself being led down the alleyway of never ending 'ifs' and 'buts' where success is simply an illusion and consolation a mere phantasm. I recall what Diane had said immediately after her return from that school in Hosuru, *'I sense an invisible hand stretching itself to strangulate the human neck.'* Am I right, Diane?"

"You are," accentuated Diane.

Rishi gave a half-hearted disturbed laugh that revealed his troubled mind. His finely tuned intuition hinted at a huge disaster in the making.

CHAPTER XI

The Institute, located in a secluded area on one of the highest terrains on the globe, was a huge landmass almost the size of a tiny village. With no town or city within a sphere of 30 kilometres, the choice of location was deliberate - to minimise the disturbances and nonconformities in the atmosphere, so that the light reaching the single aperture radio telescope, that was housed in it, could travel through less air and consequently produce images of a higher resolution.

The telescope was the size of over 20 football fields and the scope of its light gathering aperture was the largest in the world. Its motion was synchronized with the Earth's rotation to allow the camera to stare into one spot in Space, for hours. The reflector, made up of 3,250 panels, had a diameter of about 350 metres while the CCD which detected 95–97 % of the incoming light, gave the best in terms of resolution, sensitivity and just about every other parameter that an astronomer could conceive. Fainter and more remote objects which were otherwise invisible could be imaged even in very low and dim lighting.

Sridhar, an observational astronomer, considered himself lucky to have got an opportunity to work in an environment of his choice. He was a recruit; one of those bright alumni - a product of the newly established institutions that filtered prospective candidates via a stringent combination of the physical and the mental. A doctorate in Physics and Astronomy added to his credentials. The Institute hoped that, someday, he would add another dimension to its vast achievements by stumbling upon something groundbreaking that would take the world by storm. Until then, it let him be. It was an accepted credence that geniuses should never be interfered with.

To be an astronomer had been his only dream, and his ambitions had always hovered around that. Way back, when he was far younger than what he was today, the people he mingled with, including his parents, teachers, guardians and classmates, thought he was born to be a great artist or a painter. Their assumptions that he was another Picasso in the making, amused him. Of course! He could paint well and his paintings always did have the stamp of a genius, but he was too level-headed to pursue a career in the field of art, and he proved it with his impressive acquisition of a structured education.

His interest had always lain in celestial objects, and this very interest grew into a burning ambition and focussed itself into an illustrious career. Someday, he hoped to make a discovery that would take the world by storm - until then he would do what was expected of him; study the sky, the stars, the moon and other spatial illuminations.

Among many other things, his work profile, as an intern, included lesser duties like digging out images captured by the CCD and placing them in relevant folders, to be retrieved later for further study and research. It was an unimaginative job. Most people would have found it boring and monotonous. Not he. He found it stimulating. It was fun searching for something out of the ordinary in these ordinary images.

Despite his 4 years of Astronomy in the University and the Business Course that followed, his mind had not been cleared of his childhood fantasies. His teacher, who had taught him geometry in school, had impressed upon him the value of concentrating on every angle of a problem to find an optimum solution. He had taken the advice seriously. Every night was a new stage, and every day, a day of fresh surveillance. He squiggled every observation that he thought was out of the ordinary. Such jottings were hasty, impulsive and seemingly unrelated, but he was confident that, in the future, these scribblings would help him reach conclusions that would rock the world.

Probably that explained why he was considered to be so different from other college graduates who like him were functionally literate, but unlike him had very little 'sense of reality.' He was what one would term in modern slang a 'crazy weirdo,' a 'nerd.'

When the initial murmurs of the disappearance and the reappearance of the 'blinking boys' had reached his ears, he was quick to observe a correlation between the beclouding of the skies and their disappearance and reappearance. He had juggled with the idea of discussing his observations with Rishi, but restrained himself from doing so. High professional standards were demanded of him and he was unsure if his obiter dictum would be jocularly commented upon as the insight of an inexperienced mind. Prudence dictated that he wait for an opportune occasion to confirm both his observations and his thoughts, and to achieve this end he slowly but surely changed his daily schedule to include 'working hours' beyond the normal.

Work pressures in the Institute eased after 7 in the evening, but he continued to linger, on some pretext or the other. He dutifully let himself out at the appointed hour, at dusk, every working day, and strolled towards a demarcated area of the Institute where the atmosphere was least disturbed by the vagaries of nature. Perching himself on the tabled plateau which was at an altitude higher than the rest of the precincts, he looked at the sky with his naked eye. It appeared closer and clearer from here.

His qui vive had alerted him to a mathematically precise pattern. He had made a mental note of the timing of the blank images downloaded by him. His unusual presence at this spot was merely to reaffirm if such images really did always coincide with the days of sensation. The timing chosen for the exercise was deliberate, to enable a plausible justification for his observation of the interdependence between the 'changing skies,' the 'blinking boys' and the 'blank images.'

In earlier times, he had presumed that the odd faint red hue coupled with an abnormal sudden plunge into fogginess, was just a lingering effect of the setting solar body. After all, 'Sunrise' and 'Sunset' were interludes that painted the firmament with a riot of inflamed hues and shades. But now, he doubted his own inference.

He had noted that the sporadic disappearance of a young boy, and his reappearance a few days later with those 'enigmatic blinking' eyes almost always coincided with the drowning of the red shade into the unannounced misted sky. He wondered if it was synchronicity or harsh reality. Oddly, the unaccountable fogginess that clouded the air at the pertinent time, lasted only for an ephemeral time span like a punctuation. It then cleared leaving in its wake an even clearer expanse – just as if a quill had dipped itself into the inkpot to restart a new sentence with a flourish.

His mind laboured on. The incredible thing, he mentally conceded, was that it appeared not to jar as a falsehood. Some deeper questions, some intensive research, might produce some credible answers. It was too premature, though, to make an announcement. He preferred to keep his own counsel, at least for a while. He knew only too well that he was not employed as a conduit for seemingly irrelevant information, but for deducing the logical out of the extraordinary. Once in a while though, he liked to utilise his imagination.

And now as he reflected, the computer sorted through the bits of magnetic information stored in it, dug out the images that it was coded for and bounced it on to the folder specifically created for it. What he noticed astounded him. Except for a blurred patch of redness splattered across it, the CCD had been unable to capture anything on the nominated dates. It seemed as if someone out there had put a baldachin on the aperture to avoid apprehending objects beyond it. This was in stark contrast with the other dates when the pictures were crystal-clear. He scrutinised the prints microscopically and compared them, once again, with the dates given by Rishi. He had been specifically instructed to look out for inconsistencies on those specific days. He was baffled.

Picking up his yellow spiral pad in which he had been extensively making notes even before Rishi had asked for his observations, he perused the written text. It had just 3 columns – The first column had the serial number; the second, the dates and the third, the remarks. It was absurd that a man who was so technically savvy had to resort to penning down comments on a piece of paper which was considered so obsolete these days, but he did so only to convince himself.

Was it coincidence or was it just a play of his imagination? All the dates given by Rishi, without an exception, concurred with the blank red hazy images in the remarks column. It was crucial to draw Rishi's attention to it, but not tonight. Tonight, he had something more important to do; a duty he had set for himself - a metier that demanded inexorable diligence. His watch told him that it was past 9 p.m. Stepping out, he walked towards the designated expanse.

The sky was clear, and the stars twinkling. Three nights ago, he had espied an arcane curious spaceship. He had been disappointed when there had been no repeat occurrence on the nights that followed, but that did not stop him from persisting. There was a quote which said 'Fortune favours the bold' and in his case, he hoped that, tonight, it would favour the persistent. He looked up; but for a few cirri, there were no ungainly effects that marred the absolute clarity of the star-speckled empyrean firmament. He hoped against hope, as he had done during the past nights, that he would get to notice it again. All that he had to do was to keep observing. 'Patience' was the watchword.

It was sheer coincidence that Rishi, Diane and Alex chose to walk towards the parking area, at the same time. Diane was the first to notice the lone figure at a distance. She nudged Rishi, "Isn't that Sridhar? What's he doing there at this late hour?"

"Haven't the faintest idea," said Rishi. "Did tell him to look out for something abnormal - but not at this hour."

Rishi waved frantically at Sridhar in a bid to attract his attention. But there was no response.

"He's too far. A good 10 minutes of walking distance from here. I don't think he can see us from there. We can, because he is at a higher altitude," said Alex.

"Then let's go closer and find out what he is up to," prompted Diane.

They walked towards him.

"What's up, young man?" Rishi asked jocularly. "In the open today?"

Sridhar jerked. He had been so lost in contemplation that Rishi's voice came as a total surprise.

"Have been doing so for the past few days. Something's happening - seems unoffending enough except that it seems to be dovetailing with the dates given by you and that's disturbing. Just want to get the specifics on what is going on. I..."

He was cut short by Alex. "Hey!" He exclaimed, "What's that? There. Up in the sky. Am I fancying? Or is it for real?"

They all looked up at the same time.

"O yes! A strange red speck... almost blurred..." responded Diane.

"It's gone," said Rishi.

"What do you think it was? A ghost? Boy, am I glad I got to see it with the naked eye. Our images would never have revealed it. I've been told that apparitions cannot be captured on cameras." said Alex, unable to contain his enthusiasm.

"That is childhood stuff," replied Sridhar. "Ghosts don't exist. Or do they? I don't know. Seen something similar 3 nights ago but dismissed it as an illusion, because it didn't show in the images captured by the CCD. But now that you've seen it too, albeit for a fraction of a second, it's confirmed that I was not mistaken. Why is it then, that it did not reveal itself in the images? That's disturbing. Don't tell me ghosts do exist after all," stated Sridhar.

"Oh no! I was only joking. Perhaps it's an UFO?" Alex said hazarding a guess.

"An UFO? That could be a possibility," replied Sridhar, "but why are there no images?" He faltered before proceeding. "Images!" he stressed gratuitously, and laughed loudly at the word.

"What's so amusing?" queried Rishi.

"The word is an irony. This is not just a stray case. There are absolutely none even in the screenshots related to the dates given by you. All of them are a total blank except for a haze of redness. To find a sane answer, I dismissed the idealist that lay obscured in me under the guise of the intellectual, and let the down-to-earth take supersedence. I kept grappling with sheer guesswork to give a measure of authenticity to them, but it was of no use."

"Why did you not discuss it with me?" asked Rishi exasperatedly.

"I wanted to reconfirm my observation. There was a possibility that I may have been wrong. I wanted to be sure, doubly sure. It was incredulity that impelled me to walk out every single evening to search for discrepancies, and what do you think I see 3 nights ago - this strange clear speck dotting a blurred, hazy sky. When I checked the prints later, I was amazed - all I saw was an unreal opacity that focussed on nothing but a splash of weird red; the kind of stuff that's best left to science fiction. I concluded that my mind was conjuring tricks. But now that you have witnessed it, I know it's for real. I guess, our present technologies are insufficient to overcome the challenges of algorithms. 'Artificial intelligence' may have come to stay, but we are still a long way off before it reaches a peak."

All of them looked up again. There was no trace of the strange contraption; not even a faint suggestion that it may have been there only a few seconds ago.

"The lights may have been a hindrance," suggested Rishi, "Who's manning the 'Mains' today? Call the office and ask them to switch off the power at some of the terminals, and dim the area."

Alex took it upon himself to do the needful, but it turned to be in vain. With no further sightings, it was a night of disappointment. Rishi directed Sridhar to be on his toes.

* * *

Nights later, while Rishi, Diane and Alex sat at the table discussing solutions to avert the notoriety that accompanied the deaths of innocent females, they were interrupted by a call from Sridhar. He was excited, "The thing is back again in the skies." It was a clear reference to the red object that they had seen recently.

Rishi, Diane and Alex rushed to the spot. The atmosphere was hazy like the way it would have been when the particulate matter in the air reaches dangerous levels, but the object hovering and encircling near the clouds was sharper than the night before.

"It is much clearer this time. I deliberately dimmed the lights as suggested by you earlier. Sometimes the simplest of things lead us to the right solution, no matter how irrational they appear to the intellect," said Sridhar.

Rishi nodded, and Diane looked skyward, "It's receding. There, it has disappeared. How disappointing!"

"They probably sense the presence of human beings. Will they be back, do you think?" asked Alex.

"Doubtful," replied Sridhar. "Don't forget this is their third attempt this week, and as you rightly said the retreat may have something to do with our presence. It may come back, but not for some time. Anyway, I'll keep you informed."

Discouraged, Rishi, Alex and Diane turned and were about to trudge back when they were stopped on their tracks by a gasp. It was Sridhar. "I'm wrong," he exclaimed in a loud whisper. "It's back. And wow! This is even more exciting than I thought. It is much closer than before."

An 8-gon, red-coloured doodad appeared to be moving in their direction. The esoteric entity appeared like a great mechanical egret as it skimmed over the perimeter of the spot where the boys were housed. It made its way slowly towards the Institute encircling its circuitous path twice, at low altitude, before coming to a momentary standstill.

Watching it swerve round and round gracefully Alex remarked, "It sure looks like one helluva contraption. Don't recall seeing anything like it in my entire life. Odd isn't it, that it is so clear despite the atmosphere being hazy."

"Agreed," responded Diane in awe. "The only thing clear about it, is its shape. Everything else is obscure."

"I'm referring to just that - the shape. I've never seen that kind of a structure before. I need to revisit my geometry. Or maybe I'll just borrow my son's book. I may get some enlightenment there," replied Alex, enjoying the thrill of the moment so much that it brought out the child in him. "Who was it who said 'Child is the father of Man'?"

"William Wordsworth," replied Rishi. "And shut up. You are robbing us of our focus."

"Sorry!" replied Alex.

"You better be because although it is too far, it does seem as if it is zeroing in here," announced Diane.

"Oh yes! It is," said an excited Rishi. "It was the same last time wasn't it, Sridhar?"

Sridhar nodded, "Yeah, but never this close, and never this long."

"Notice something?" asked Diane. "It appears to be hovering exactly above the place where we have insulated the boys, although I can't be too sure. It is too far away for me to decipher its exact location. I sense an overt operation; an attack on the premises, perhaps. Should we take cover?"

"I think we should," said Alex in a hushed tone. "There is some red smoke being tunnelled precisely over the spot where the boys are kept. It's dangerously threatening. Is it planning to lift the boys, by any chance? I hope there's no such intent. What do we do, Sridhar?"

"Just flood the area with lights. Maybe that'll do the trick," advised Sridhar.

Alex relayed the relevant instruction. The entire zone was swamped with bright lights. The red object moved out of their line of vision, evidently back to where it belonged but not before leaving webbed streaks of red light in its trail. The sky cleared visibly the moment it disappeared.

"What or who do you think it was?' asked Alex rather incredulously. "What could be the motive?"

"That is yet to be determined. In the meanwhile, check on the boys, Alex," said Rishi. He then turned towards Sridhar and asked him to scrutinise the downloaded images relating to the 'moment' and send the prints to his office. And then in an offhand manner told Diane, "I think we should send the pictures to Ayaz for reviewing. My intuition tells me that this would probably be an eye-opener to a vast amalgam of possibilities. I may sound insane, but I think I'm not wrong in assuming that this moment and the images related to it, may offer solutions to the incredulous mystery of the 'blinking boys.' Remember the thing kept lingering just there. It must have sensed a connection."

***

was already waiting in the wings as a replacement, he might as well lend an ear to it.

"You surprise me. What is it?"

"The fundamental design of the microchip will remain untouched," illustrated Pontus. "There'll only be a slight tweaking to change the seat of incursion. In the original, it was the 'eyes'; in the modification, it will be the 'nose'. The redesigned chip will be far more effective. The only thing that will not change is the tedious process of picking up human samples. Initially one, then two and then many more after the preliminary ones have been successfully tested for their capability in terms of the precision and calculation in its tangible form, but that can be handled with kid's gloves. Besides, there may not be a need for too many samples in this case."

"Why not?" queried Maxus.

"It has been designed to have an exponential effect," asserted Pontus.

"So, we do have a solution, after all," Maxus sounded relieved.

"Yes," confirmed Pontus.

"But then, what about the instance of 3 signals from one single location? We have an explanation for their nestling, but do we have a solution to counter it. We were to airlift the samples. Why have we not done that? Logistical problems?" asked Maxus

"No," replied Pontus. "No logistical problems. The region had been recced comprehensively before embarking, so there were no snags or glitches during the trip. It was smooth. In this case, the Earthlings, themselves, were the biggest hurdle."

Maxus gave him an acerbic look.

Pontus sobered. He was au courant with the repugnant confrontation, leaving no detail, "The spaceship didn't fail. Despite the challenge of gravity, it thrust into the specified destination. The intention was clear; to pick them, study them for any fiddling in the spaceship itself, and then scatter them ... each several Earth kilometres from one another. Just when we thought we had touched success, the bright artificial lights threw a spanner in the works. In an instant, the aerial zone was floodlit, forcing us to swerve off the tracks. The area is apparently well guarded even at that hour."

Maxus expressed displeasure, "A matter of a few Space moments, you say. You should have forged ahead. Can't imagine why you retreated! It was never the case in the past. We had always succeeded in smuggling out a human clandestinely, avoiding detection even in the most populated locales. What made it so different and difficult this time?"

Pontus appeared exasperated. He attempted to explain his stance in as calm a manner as possible, "Picking up a human sample, one at a time, is relatively easier. Our previous visits entailed just that. Such operations warrant the use of a 'Mono capsule' which has a unique feature of self-releasing atmospheric gas that acts like a smokescreen, while travelling through

Space and while on Earth. The only telltale sign is the misting of the troposphere and a slight pigmentation, both of which can be easily attributed to the pollution levels and the colours of the Sun. The occasions when we confront some difficulty, and that too not a technical one, are when we have to pervade overpeopled areas. In such places, we have to disguise ourselves to stay off the radar of the suspicions of the Earthlings."

"The same could have been repeated here," counselled Maxus.

"That would have defeated our purpose," replied Pontus.

There ensued a strangled silence, at the end of which Maxus said, "I fail to see why."

"Picking up just one sample and leaving the other 2 behind would have defeated our rationale," enumerated Pontus. "It's an established given that the 3 samples have been assembled there for a motive - why else would they be there? That the chip has not been dabbled with, is also well established - the signals say so. What is not established is how much is known and what."

Pontus looked at Maxus for some verbal response. There was none. Only a cynical look expecting a more elaborate answer.

"One missing sample, out of the 3, would have fuelled their suspicions," Pontus continued. "'Whodunnit?' That's the question that would naturally follow. And the consequences!!!... The consequences would be just what we wish to avoid ... Beefed up security... round the clock vigilance, continual experimentation, more snoopings and what have you. And then would follow the short listing of suspects. Under such circumspect conditions, picking up the other 2 would have been highly improbable."

"3 'Mono capsules' could have been used one after another in succession," derided Maxus.

"It has probably escaped your memory that the 'Mono capsule' is not only self destructing, but also route-sensitive," remarked Pontus. "Once the 'pick up' operation is complete, it is kept in 'use' mode for the next few 'Earth' days to enable transportation of the implanted human back to Earth. It then 'autopilots' its way back on the same route only to fizzle, leaving no traces of its existence except for an opaque unnavigable trail making it impossible for another similar 'Capsule' to follow the same track. That is one of the many reasons why we have never iterated the same place twice for picking up samples. They have customarily been picked up from places far flung from one another."

The assessment was grim, but it was the truth. Maxus had to accept it.

"Haven't we been trying to resolve this problem of the 'Mono capsule' for eons now?" Maxus asked. "It's an extraordinary invention with several competences. It's a pity its usage is limited. I was under the impression that your field experience would have facilitated us in finding a solution. Clearly, that is not the case. If we are nowhere close to resolving it, then it's time we initiated substitutions. Any project that takes an uncharacteristically long time to reach a closure should be abandoned. We are regulated to accomplish the impossible and if we fail to accomplish it within a deadline, the conclusion is clear - it can't be accomplished at all."

Pontus let muteness dominate the conversation before replying resignedly, "I know the sitch is disillusioning, but we will have to be content with it, at least for now. A solution's being worked out and hopefully something will have been achieved soon. Until then our 'Mono capsule' stays."

"Hopefully," grumbled Maxus in a disgruntled tone. "But I still persist that retreating at the nth hour was not advisable. You were so close, so near. You could have gone ahead and just picked them all up."

"It is not as simple," opined Pontus. "Orchestrating an operation of this dimension opens doors for the Earthlings to cast their suspicions on the extraterrestrial. Sagacity dictates that we let the Earthlings believe that the onus of blame lies entirely on their shoulders. It is the cardinal tenet of Martian standards to leave no obvious clues behind."

"Clues?" snorted Maxus. "What clues? Are we the only ones to inhabit the Cosmos? It is filled with countless entities. Anything out there could be responsible."

"The reasoning, admittedly, is sensible... but not pragmatic."

"Not pragmatic?" scoffed Maxus.

"Yes, not pragmatic," affirmed Pontus. "We are the closest; reason enough for the needle of suspicion to be pointed towards us. Their obdurate persistence to experiment with Space and discover living creatures in fresher pastures has led them to develop quite a few technological thingamajigs. You remember the weird thing that landed on one of our dead craters? They call it a satellite, an orbiter or rover...? Anyway what's in a name? It is evident that it has collected quite a few samples from there."

"How would that pose a problem?" queried Maxus

"In many ways, or so it seems to me," expounded Pontus. "Those samples will give them plenty of room to conclude that the spaceship hovering overhead actually belongs to us."

To Maxus, this was like cold water bespattered on a raging fire. He stuttered, "I doubt."

"I'm stating a possibility," said Pontus. "Appearances are deceptive. They are not as guileless as they appear to be. Their interest in our planet is dangerously tilted to suit their ambitions. Either they are investigating the existence of life here, or it could be something more imperilling."

"Something more imperilling?" asked Maxus, showing sudden interest.

"Yes," uttered Pontus. "Like us, they may be seeking an excuse to conquer us. I must admit, though, that they are living a Utopian dream. Their tenacity and intellectual doggedness does astonish me but these admirable qualities stand negated against their lack of team spirit. Their one upmanship is their greatest undoing. They have this monstrously queer desire to outdo each other in competitive skills. That's what makes things so difficult for them."

"Difficult?" intoned Maxus. "That's an understatement. You are stating just the mental facet... Have you considered the physical? Would they be able to sustain themselves in 'water and oxygen restricted' conditions?"

"No," underlined Pontus.

"Do they possess spacecraft that travel at several times the speed of light like we do?"

"No."

Maxus stopped short of asking any further questions. He then went on to say, "Now both you and I know the homework involved in restructuring a situation to keep the humans alive during their transportation as well as their stay. All the parameters are delicately considered while they are in residence here... even their skin which gets thinner while traversing through Space. Thanks to our research, study, analysis, experimentation and dummy runs, we now have this cosmetic gel that preserves it in its original form. Do they have that know-how?"

"No," accentuated Pontus.

"Do they have technology that can shield them from the adverse effects of the heavy ion radiation?"

"No! No! No!" repeated Pontus.

"So, clearly our fears are misplaced," accentuated Maxus.

Pontus replied in a deliberate manner, as if weighing his words, "I disagree. I mean I disagree with your view that our fears are misplaced. They are not the kind to be daunted by technological deterrents. They work on presumptions that have all the artifice of brainpower and are cruising along the right path. They are bound to find solutions soon."

"I have my qualms," expressed Maxus.

"Did we know, all that we know now, in the immediate past?" asked Pontus.

"No," said Maxus shakily.

"We learnt. They will too," Pontus asserted.

"I think we are erring too much on the side of caution when we start thinking on such lines," stated Maxus explicitly.

"I wish I could agree with you, but I don't," said Pontus. "Since nothing is visible superficially on our planet, I shan't be surprised if they propel giant sensors and electronic prongs next, to see if they can find liquid water several Martian miles below the surface. Enough cause for worry."

"These are frivolous distractions that offer no serious threat to us," countered Maxus.

"What you term as 'frivolous distractions' are not as frivolous as you state. They can become serious constraints if they occur once too often," argued Pontus. "Being presumptuous is a dangerous attitude. Their constant intrusions will put a strain on our time and energy, and distract us from our prime focus of establishing supremacy in Space. That is why it was pertinent for us to make an immediate getaway. When things quieten, we shall execute a second, hopefully, unblemished attempt to airlift all the 3 specimens that are located there."

"If you think you are correct, we can rest that for a while," relented Maxus, "and get back to discussing the modified design of the chip. You had discussed a 'substitute' in lieu of the original. When do we introduce that?"

"Once the case of the missing sample has been solved," declared Pontus.

"Is that a compulsion?" asked Maxus.

"Yes," contended Pontus. "The problem is titanic and with no obvious indications to suspect the Earthlings, the setback has to be compulsorily tackled before we move forward. After all, it'll have to be factored into our new design to ensure that it doesn't reoccur. Untraceable missing samples imply 'missing links' and 'missing links' indicate invisible nemeses lurking in the fringes. They have to be found and nipped in the bud before they turn into uncontrollable monsters."

"But this 'provisional suspension' doesn't augur well for the smooth flow of our Plans," countered a sullen Maxus.

"The suspension is impermanent," justified Pontus. "It's not a debacle, only an impediment, but one that can rear its ugly head if not tackled at the appropriate moment. I wouldn't hesitate to add that obstacles like these are the 'core' of Martian success. Such holdups only help in perfecting our skills further and making us more resolute on our path to victory. An alert has been sounded; the antennae switched. It may entail a few Martian days before we reach a plausible inference regarding the sample. A suitable measure can then be adopted."

"Then let's act without wasting time," said Maxus. "In the wider scheme of things the suspicions of the Earthlings are of little relevance to us. It's the Venusians that are a cause for concern. The more I think, the more I suspect them. They are Machiavellian. We must stop them from smelling a rat, that is, if they have already not."

✳ ✳ ✳

CHAPTER XIII

Fitted with high definition cosmic cameras, their 'Trace Trajectory System' had the capacity and the capability of recording movements over a trillion Venusian miles away. Immaculate in conception and faultless in accomplishment, it had been propelled into Space to monitor the activities of the Martians. The deliverables were enormous. That the Martians were once again working on a gambit to dominate 'Space' was abundantly clear. It was evident that they culled human specimens and deposited them back exactly at the spot or somewhere close to it days later, but what happened to them while they remained imprisoned with them, was not clear. The Martian fortress of clandestine activities was impregnable and the Venusians were completely alienated from the ongoings within the 'Red Environs.'

It was neither prudent nor fitting for them to base their conclusion on an extensive cache of suppositions and circumstantial evidence. Their surmises had to be reaffirmed with something concrete and substantial.

It was Athenia who hit upon a brainwave. She suggested that they pick up 2 individuals simultaneously - one touched by the Martians and the other not, and then study them for differences. It was brilliant theoretically, but practically it lacked the fluidity so indispensable for the kind of success envisaged by them. Fundamental factors such as 'timing,' 'anonymity,' 'technicality,' 'navigability' and 'cosmic constraints,' had to all coalesce if it had to be accomplished.

The task was overwhelmingly challenging and intricate. It was nigh impossible to say the least, but that did not deter them. Their grit and determination to circumvent hurdles and overcome obstacles stood in their favour. To add to it, they had the relevant knowledge and the corresponding brainpower to implement it. As a result they were now staring down at the 2 bodies that lay before them wondering if the exercise had really been worthwhile. Externally they appeared similar. There was nothing in their physiognomy to suggest that any one of them, in any way, was different from the other.

Athenia glanced at Phoebia ruefully. "Do you think our decision was wise?"

Phoebia gave no reply. She was involved in gawping at the bodies lying on the solidified gas slab, curiously.

"You seem to be in awe of them. Have you never seen them before?" Athenia said, in an attempt to shake her from her musing.

"No, never." replied Phoebia. "Never had the opportunity. You of course did see them when you spearheaded the travel to Earth during our last mission."

"Yes. I did," agreed Athenia.

"Eccentric creatures... but Daedalian, I must admit." Phoebia remarked.

"Yes!" granted Athenia "Too higgledy piggledy for my liking, but we can have a confab on them later. Right now they are here to be analysed for traces of the Martian element and not as extraterrestrial objects to be studied. We have to consolidate on the opportunity and engross in the task that we have set for ourselves."

Phoebia gave a brief nod. Athenia was right in mentioning that there were no outward signs of meddling, but the longer she looked at them, the more her extrasensory perception intuited that they would unearth some dissimilarity, if not externally than probably internally.

"I totally subscribe to that," she replied in a daze that she had yet not jiggled herself from. "It's discouraging to note that there's no indication of tampering, at least not superficially. But the possibility that the Martian guest may not have been dabbled with, is remote. There can be no other reason for the lifting of the body and housing it for days, and finally putting it back to where it belongs. Nobody, in their right senses, will incarcerate an entity, living or otherwise, unless they have a valid reason. The disparity may reveal itself only after a comprehensive and meticulous assessment."

"Are you insinuating that the bodies will have to be subjected to a thorough examination and evaluation?" asked Athenia, her discomposure writ large over her countenance.

"Yes," relented Phoebia.

"It's parlous," shrieked Athenia. "I want no harm to come to them. That's not why we towed them into our territory."

"Having brought them this far, amidst stringent checks, it would be sheer folly not to do so," reasoned Phoebia. "No harm will come to them. It'll entail but a few Venusian moments for the 'Goliath' screen to scrutinise discrepancies, if any, between the two. It'll be done even before we begin."

The 'Venus XXX Scanner' and the 'Goliath Screen' made for a perfect, precise pair. They were placed alongside each other, for easy viewing. As the bodies went under the scanner, the screen magnified details multiple times over, to enable a lucid observation of the most nanoscopic feature. The Scanner slid seamlessly, phase wise, halting at every feature, from the top of the head, through the forehead, the eyebrows, the eyes, the eye sockets, the nose, the cheeks, and the mouth but detected no variants, no divergences, no differences, that is, not until they reached the part of the human ears that touched the rear end of the skull.

"Halt!" enjoined Phoebia quite startled by what she thought she saw. "There, can you see a radiolucent metallic chip just a little below the right ear? If I'm not mistaken, that's the culprit. It is incongruent with the rest of the elements of the human body and is visible only in the Martian guest."

Athenia peered with greater concentration and focus. "Goodness gracious! You are right. It is so infinitesimal that it literally blinked past my scrutiny.

But for you, I would have missed it. You are unduly canny. And you prove me right, again and again, always."

Phoebia bowed her head in a show of modesty and then said what she had to say, "Yes, I agree it is inappreciable, but that's what makes it so compelling."

"And now to ask an idiotic question that is asinine because it was I who accused the Martians for the misdeeds on Earth, and it is strange that it is I who should be asking you, but it's important that I do," Athenia said.

"Go ahead," prompted Phoebia.

"How do we know who put it there? Or do we have to find that out too?" enquired Athenia.

"No, we don't," replied Phoebia with the easy assurance of an expert who knows her subject only too well. "It's easy. A situation compounded by the Martians can never remain undetected for long. Entirely blinded by self-interest, they leave no stone unturned in putting their signature, in everything they do. Can you see the red blur all around it? That is the Martian stamp."

"So, I am right, after all," Athenia claimed with a tinge of pride.

"Yes, you are," consented Phoebia. "But we can't rest on this finding, alone. The chip will have to be decrypted to understand the schema. It indubitably endorses an intricate hieroglyphic."

"And pray, how would we do that?"

After an instant of barely perceptible pause, Phoebia replied, "With our Venusian micro needle."

"The micro needle?" Athenia vented "With its mortiferous drops? Oh No! The needle and the drops make for a noxious combination. One of our most lethal inventions! We were elated when we contrived it, only to rue later. We have vowed never to employ it, but in exceptional circumstances, and that too as an ultimate resort. If we are not careful, the corollaries could be perilous. It would be hazardous to use it on these susceptible creatures. In our zeal to be the protectors, we should not end up being the perpetrators."

Athenia was distressed, very distressed.

"I'm in sync with you on that," said Phoebia pressing her argument. "Like you, I too am aware that the Earthlings are not as impervious as we, but this is an unusual crisis that calls for unconventional solutions. The Martians have to be reined in before it is too late."

"The idea doesn't appeal to me," persisted Athenia.

"I reiterate that we have no reason for worry," assured Phoebia. "No doubt, it's a dexterous operation, but my knowledge will manoeuvre it to success. The dosage will ensure that. A drop of the numbing anaesthetic liquid is the ballpark for procedures of this type. Administering it exactly on the spot will do the trick."

"You sound so confident," appreciated Athenia.

"I am," contended Phoebia. "But yes, one wrong move and the consequences will not be to our liking.

An optimum estimate will heal; an increase or decrease, even the slightest, would result in devastating ramifications. Admittedly it's a nimble-fingered procedure that demands maximum precaution to avoid inflecting injury, and adopt obligatory processes to eschew any pain or impairment. We have to be strictly vigilant while the needlework is on."

"And the after-effects? What about the after-effects? It is not enough to certify their safety during the procedure; we have to consider the repercussions," warned Athenia. "There is bound to be some adverse effect which may impact them negatively. A state of neurosis? A state of unconsciousness? A state of stupor? A state of physical disorder? A state of abnormality to their senses? Now don't try to paint a rosy picture by saying that there aren't any. I would refuse to believe you, even if you do say so to put my mind at ease. Every action does have some reaction, and the same would apply here."

Phoebia looked at her smugly with a ghost of a smile on her face.

"I know, I know," Athenia continued, "I sound like a novice when I ask such questions, but you can't deny that I share a close bond with them since I am the one who is always involved in their causes. Besides, as you so often say, I am a sentimental fool; neither a pragmatist nor a strategist - only an observer and a bit player in the entire chessboard of the Venusian Cosmos."

"You misunderstand me," replied Phoebia. "When I censure your actions with such statements, it is neither your reasoning nor your rationale that is under a cloud. Such moderations are alerts to caution you into keeping a lid over your emotions. Misplaced fervour can overshadow logic, leading to biased deductions that can have ruinous outcomes."

"I understand your concern, though I don't necessarily agree with them," said Athenia. "I refuse to believe that the consequence of this surgical manoeuvre will be anything but deleterious. I recall the moment when the cited research was profiled for approval. It had been thoroughly dissected both for its weaknesses and strengths, and a conscious decision was taken to keep it in abeyance until it was further improved upon."

"I remember too," maintained Phoebia. "But I also distinctly remember adding a caveat - a proviso permitting us to use it in situations where pros outweigh cons."

"Reconsider! Please do. I know I'm being obstinate when I continue with my insistence," persisted Athenia. "But I'm in no mood for cataclysmic upshots."

"There you go again!" smirked Phoebia. "Rest assured no harm will come to them."

"Them? I thought just one has to be surgically tinkered with - the one who's housing the chip," said a mystified Athenia.

"No," stated Phoebia firmly. "Both will have to be subjected to the procedure. It's a precautionary measure. We want no situation where we may be constrained to return to Earth merely to lug them back, for further testing. It's of overriding importance to remove the chip from the body that's sheltering it, and validate that the other is free of it, despite the fact that there are no indications of its presence in it."

Athenia sounded hysterical, "Two of them at the mercy of our atrocities! Now how do I protect them from you?"

"You don't have to," reassured Phoebia. "The outcome will not be that of destruction. On the contrary, the postscript will be a pleasant one. They will regain consciousness, a few hours after they have been transported back to Earth, and behave as normally as any other living being out there, at least physically."

"There, I said so. There may not be any hiccup physically, but mentally? Emotionally?" intervened Athenia, a little impatiently.

"Hear me out, will you?" Phobia cried. "Your fears can be expressed once I've done with my explanation. The only anomalous factor will be the emission of a pale bluish tinge from the spot that has been operated upon; a nuance that will be visible only in total darkness. During the day, the glare of the sunlight will dwarf it completely. And there'll be one other apparent effect in the broad spectrum matrix that will be evident to one and all on Earth, irrespective of day or night."

"And what would that be?" asked Athenia.

Phoebia looked at her patronisingly before replying, "A temporary healing effect that will be dominant in the initial days, only to peter with each passing day until it totally ceases to be. The individuals themselves will suffer from no uneasiness, but they will most certainly exude a calmness that will soothe the nerves of those they come across."

"You mean, they'll be blessed with an unnatural therapeutic power?" Athenia said.

"Yes," confirmed Phoebia. "The anaesthetic will suppress emotions of aggression of any brain sedated by it, while upholding the peaceful ones. As a result, when the respective human locks eye with another, the beneficiary will feel at peace; nerves will be calmed - and aches and pains if any, will be gone."

"You amaze me. That's interesting. Why was this never told?" demanded Athenia.

"For fear that it could be exploited and misused," stated Phoebia blandly.

"Is that an accusation? I know I get emotional in fits and starts, but I would never compromise with our principles," said Athenia, sounding hurt by the statement.

"No, it's not an accusation, but if the cap suits you, you may wear it. But... But I must state very emphatically, that it's the dosage that decides the result; we have to be extremely mindful of that. Our carelessness would ruin it all," reiterated Phoebia. "But..." Phoebia hesitated.

"But... what?" coaxed Athenia.

"But this curative effect will be temporary," declared Phoebia. "It will last until such time that the referred blue tinge dissipates totally. With the fading of the incandescence, so will their healing power. On rough calculation that would be approximately 3 Earth months; maybe lesser, but certainly not more."

"Only 3 months? What leads you to conclude that it will last for only 3 Earth months?" asked a resentful Athenia.

Phoebia's earlier statement had troubled her greatly, but she couldn't let that interfere with the work in hand. She had to let bygones be bygones if they had to work in unison for the betterment of the Cosmos.

"The dosage," articulated Phoebia. "It's a constructed and precise calculation. In this case, for instance, we have to dispense with the minimal measure to achieve the desired outcome, and the outcome of such a miniscule portion lasts for just that period. With the dispelling of the radiance, so will the degree of healing. They will then return to what they always were."

Athenia appeared satisfied with the justification. Being a utilitarian, she had little or no pedagogical knowledge. Academics were Phoebia's prerogative. She merely worked relentlessly to get the coveted results by executing postulates and concocting strategies based on theories propounded by Phoebia.

* * *

The chip, when dislodged, generated a delirious zeal. Athenia was awestruck. It was a beauty, so seemingly inoffensive that it could dupe anyone into assuming that it was incapable of anything evil. But Phoebia knew better. It had to have some interconnection with the human body. Why else would it be put there? There had to be a pattern of impulses which decoded itself into a reoriented motif, the instant the body landed on Earth. It was important to study it in detail if its involvedness had to be comprehended. She put the chip under the 'Cosmic Max' magnifier and waited for it to be sorted through the bits of magnetic information encrypted into it.

The Martians had not changed at all, and probably never would. How could the Venusians have been fooled into thinking that they had? On the contrary, they were acting on their ambitions with greater zeal and resolve now than before - reason enough to cause apprehensions and provoke prioritized action. Their latest act only served to confirm that. They were doing exhaustive research on the Earthlings - to the point of obsession. What else could explain their tactic which was an art by itself?

Phoebia pondered at the size of the data bank of the Martians and wondered, in alarm, if they had a repository of data on their planet too. Not that assessing information on their terrain was easy; Venus was characterised by such high temperatures that approaching it would be almost impossible for any other cosmic entities. They would go combust much before they had even reached the gravitational field. The Venusians had a mercurial persona with the propensity to withstand high degrees of heat in their own planet, as well as tolerate the temperatures on Earth and Mars.

But the Martians were a disparate breed with an insatiable ambition to exercise sovereignty over the 'Cosmos.' They would do anything to surmount even the blazing heat of their planet. Colonizing Earth was the first step in reifying this ambition.

The mechanics of the Martians' 'Plan' was never in doubt and that was what had prompted the Venusians to intervene on a previous occasion. At that time the approach had been different. The Martians had opted for a direct attack on 'Earth' by releasing a deadly 'virus' to cause an epidemic. The repercussions would have been colossal had they succeeded. The Earth would have been wiped off of all the living species leaving it exposed for the Martians to enter.

In the combat that followed, the Martians and the Venusians had been an equal match; but the former had been handicapped by the scorching heat that tilted the victory in favour of the Venusians.

The defeat was a warning to the Martians. The Earth had an ally, and they an adversary - an adversary who was as well equipped as they, if not better... an adversary who was not as naïve as the Earthlings... and above all an adversary who would not think twice before directing their long-range, fine-tipped missiles towards them, the denouement of which would be the destruction of Mars.

The Martians had retreated in a show of peace, not due to ineptitude but because of force of circumstances. In a 'Cosmic War' of such magnitude 'Earth' would have been the biggest casualty, and they were shrewd enough to comprehend that. They wished to retain both - their planet as well as 'Earth.' The Law of Balances had to be ensured, if Space was to retain its 'status quo'.

Phoebia reviewed the chronology of events leading to that particular *'spatial confrontation'* and its termination. She posited if the Martians had relegated it as an anachronism which would have no sequel in the future or if they were basking in the presumption that the Venusians, in their preoccupation with other cosmic matters, may have disregarded it. Knowing their temerity, the latter seemed more likely. What else would explain their choice of a subtle tactic, this time, to wipe out the human species? The game plan was immaculate - picking just one individual, at a time - but all targeted for a single purpose.

She experienced a moment of hesitancy and theorized if she was prudent in her obtrusion. After all, what did it matter if they did conquer Earth? The Earthlings didn't care. Why should she? Deep introspection on the subject gave clarity to her act. With virtually 2 planets under their belt, the Martians would be Masters of Space, but would that satiate their greed? No. She was well acquainted with their craftiness to infer that once Earth was conquered, 'Venus' would be their next target.

Her stance radiated a feverish urgency. She had to shroud Earth with a mantle of protection, and she knew why. She was selfish. Earth was in closer proximity to Venus. Conquest of Earth would make it far easier for the Martians to negotiate victory over them because of the shorter distance. As for the heat, she knew their selective intellectual prowess would find a way out. The thought rocked her to an abrupt adrenal alertness and staunched her resoluteness. It was imperative to abort all their attempts. She kept her reflections to herself and deluded the others into thinking that 'peace' in Cosmos' was her only intention.

Her diligence had not failed to take cognizance of the reciprocal relationship between 'Mars' and 'Earth.' Like the Martians, whose unwarranted attention on Earth were irrefutable, so was the tremendous interest in Mars, by 'Earth'; the dissimilarity lay in their intentions. One was destructive while the other was not. The Earthlings were non-combative or so she chose to believe. They were certainly better trusted.

She returned her attention to the task in hand - the examination of the chip. The miniscule chip had to be enhanced manifold even under their specially designed 'Cosmic Max Amplifier.' As she intricately decoded the nomenclature whose core deliverability was the death of an inoffensive female human being while protecting the male, her frenzy reached a crescendo. What she stumbled upon surpassed her wildest imagination. Ingenuity or Genius? What would be the appurtenant description? Truly the Venusians had misjudged the Martians' acuity. She grudgingly credited them for their detailed study on human beings. They had overlooked nothing - nothing at all.

CHAPTER XIV

The gargantuan expansive 'Visual Display Unit' displayed the findings. Phoebia explained the interface, the cue, the direction and the execution of what she had discovered. "And yes," she said to Athenia, "The chip precisely duplicates our presentiment. The trademark shade of red which was not too distinct at the time of scanning is not illusory - it is real, reconfirming our darkest suspicions. The Martian's involvement as presumed by you before the scanning and as adjudged by us after it, has proved to be accurate. They are the villains. The chip is a magnificent piece of digital architecture modulated to avoid recognition. The entire operation involves an impeccable application of erudition and a high degree of commitment."

"Erudition?" asked a surprised Athenia.

"Yes, erudition," confirmed Phoebia.

"I figured it as much," decreed Athenia. "The casualties are so well orchestrated that they appear to be a part of a finely coded synchronization... Simple if you know the menu... Befuddled if you don't. But Martians and erudition? That's a bit of contradiction, isn't it? Your statement has triggered my sense of the unknown. I always presupposed that they were more aggression than intellect."

"Frankly, I could never have credited them with so much cerebration!" said Phoebia. "Wrapped in ambiguity, it's an 'action program' way beyond the cogitable and has been ingeniously designed to earmark the 23rd pair of the human chromosome."

"Doesn't make any sense to me," announced Athenia.

"It wouldn't, not until you have some basic knowledge on the human cell," replied Phoebia.

"The human cell?" asked Athenia. "I'm afraid, that doesn't make any sense either. My mastery on the subject is limited. You'll have to enlighten me further."

Phoebia condensed her tutelage to the bare specifics, "The human body consists of several thousands of human cells, in each of which is a molecule called the DNA or deoxyribonucleic acid. This DNA is coiled tightly many times around something called 'proteins' in structures called the chromosomes. These chromosomes are visible in the nucleus of the human cell only at the time of cell division and not otherwise. The human beings, themselves, have learnt about them by observing them during cell division."

"Remarkable!" exclaimed Athenia as she geared herself to give maximum attention to something so intriguing that it appeared to be a marvel by itself. "Now that's what I call a mystery! Imagine being visible only at the time of cell division and not before or after that! Peculiar isn't it?"

"Yes," said Phoebia crisply. "My limited research indicates that at the time of cell division, these chromosomes become highly condensed and are then visible as dark distinct bodies."

"Now, what could be the explanation for that?" asked Athenia evidently intrigued at the behaviour of this all important but so tiny a thing called the 'cell.'

Phoebia admitted her ignorance on that aspect, "I wish I knew more, but I don't. Our studies have a strong base on cosmic information and logic. Sadly, human biology is neither an extensive part of it nor a necessity. We have no need for it. It suffices us to know that the human body is complex and as said earlier made up of cells."

"This is puzzling! If the humans can observe such a microscopic cell under the microscope, why have they failed to detect the 'Martian Chip'?" Athenia asked.

"That is the beauty of this Martian Chip," proclaimed Phoebia. "We can't deny acclaim for their inventiveness. The size is too infinitesimal to be detected and determined by the limited sources available on Earth. It is even smaller than the human cell. Things are easier on our Planet. Our 'Cosmic Max Amplifier' can magnify things billion times more than a human microscope can. It is no wonder then that we managed to detect it. Another thousand years; and the Earthlings will still be struggling with an innovation of this kind. Anyway, their technical finesse or rather the lack of it, is of no relevance to us. Their chromosomes are."

"Oh yes, understanding the chromosomes is ad rem in the light of understanding the methodology of the Martians," agreed Athenia.

"Every human cell normally has 23 pairs of chromosomes, 22 of which look alike in both females and males, while the 23rd pair is distinctly different. This last pair referred to as the sex chromosomes is responsible for the gender of the human child when it is born," Phoebia continued.

"Wow!" exclaimed Athenia.

"The DNA is a double helix formed by base pairs attached to a sugar phosphate backbone. It can make copies of itself, and is the hereditary material in humans and almost all other living organisms on Earth. The sequence of these base pairs is exactly like the set of codes that we have in our Space laboratory," elaborated Phoebia.

"Too abstruse for me to grasp," claimed Athenia.

"Let me simplify. We have been developing various codes over the years and placing each of these codes singularly and separately in independent vaults. Why do we do it?" asked Phoebia.

"To pick specific codes, assort them and then arrange them in a predetermined form to get the desired result, every time we wish to experiment with a new activity in Space," answered Athenia.

"Absolutely," affirmed Phoebia. "So is it with the human beings. The only difference lies in the manifestation. In human beings, the mutations occur naturally to beget variations, while we choose to do it mindfully."

"That explains the difference in colours, heights, weights and every diminutive characteristic of these human beings," announced Athenia.

"Correctly said," conceded Phoebia.

"You know what the Earthlings would call themselves, if they had to see things from our point of view? An assorted box of chocolates," chuckled Athenia.

"Chocolates?" It was Phoebia's turn to plead ignorance now.

"It's an edible that's available there," clarified Athenia. "Now, don't ask me for an exposition. You need to be on the field to know what I mean. You may be a theorist with enough knowledge to last an epoch, but when it comes to an acutely tuned cosmic sense and an awareness of how to apply it, I can beat you hollow. But in all fairness to you, your knowledge is the base on which I have applied techniques to get tangible results. I can tell the 'what' and 'who' of a cosmic situation but not necessarily the 'why.'"

"I have never ever denied that," acknowledged Phoebia. "That is why we need to be in tune all the while. But to revert to what I was saying, the sex chromosomes 'X' and 'Y' determine the sex of a human individual. The males have one 'X' and one 'Y' chromosome while the females have 2 'X' chromosomes."

"O Heavens! Are human beings really so jumbled?" said Venusian II, who until now along with the others, was listening with rapt attention to the dialogue between the 2 most prominent representatives of their Planet.

"O Yes! They are. What is bewildering, though, is that the Martians have been able to study all this and have skilfully used it to suit their evil designs," responded Phoebia.

"I'm dumbfounded," said a surprised Athenia.

"So am I," conceded Phoebia. "It is inconceivable how they have pervaded right up to the genetic makeup of the human DNA. They have actually established a mechanism by which electronic impulses from the chip are transferred, through a complex grid of micro infinitesimal connections, to orchestrate them to target only the 'XX' chromosome of the DNA. A bed of microprocessors is then permanently embedded on the telomeres of the recipient's chromosomes. These processors are then sensitised to direct impulses to the immune system of the targeted human being. The result is a slow but sure termination of the life of the individual."

"How clever!" Athenia exclaimed.

"Yes. Very clever," concurred Phoebia. "The microchip is designed to send out a stinging ray, at lightning speed, towards the 23rd pair of the chromosome. The ray strikes only if it senses 2 'X' chromosomes. An encounter with even a single Y chromosome acts as a deterrent redirecting the impulse to its origin and rendering it ineffective. It moves with such alacrity that an ordinary human eye will find it difficult to detect, unless observed with great intent, and even then, it would wonder if it had seen it at all."

"Impressive!" said a startled Athenia. She took a moment to compose herself before continuing her conversation. "I wonder why the Earthlings failed to notice what we just have. I know, I know, you did say that an ordinary human eye will fail to register it; but all are not ordinary out there, or are they? Or have they chosen to ignore the obvious? Lack of advance techniques may have made it onerous, for them, to detect the chip, but the streak.....that should have been visible. All it requires is an alert mind and an eagle eye."

"Don't underrate them. I think it was you who informed me that they have invented an incongruous widget that doubles up for a unique pair of eyeglasses. You also added that most of the affected human beings are sporting them. If it is true, I am certain they may have detected this," countered Phoebia.

"You think so?" asked Athenia expectantly.

"Yes, I do," replied Phoebia. "Their technical competence may not be strong enough to detect a microchip as tiny as this, but a few brighter ones may have noticed the red streak and connected it with the inexplicable 'blinking feature' of the young males. The eyeglasses may merely be a superficial solution to bide over what they think is a temporary problem. The root cause evades them and always will, until such time that they advance technically to be on par with other entities of Space, like us."

"Are you sure?" queried Athenia.

"Yes," said Phoebia emphatically.

"I concede that their technological competency is not robust enough to detect a microchip as tiny as this, not because I believe in it, but because you repeatedly keep telling me so. You give me no occasion to forget it, every time we encounter a problem related to Earth" answered Athenia, "but that may not be the sole reason for their laxity in finding the source of the problem."

"What other reason can there be?" asked Phoebia with a puzzled look.

"Their lack of knowledge of alien existence. That is their biggest handicap," explained Athenia. "Naïve in their conviction that no life exists beyond their borders, they'll snoop around for the source of the pandemic within their planet, and not outside. For them Mars, Venus and other planets are barren and devoid of inhabitation."

Phoebia disagreed, "I'm disinclined to agree with you, here. Their actions prove otherwise. Haven't they only recently started propelling 'objects' into outer Space more specifically on 'Mars'? There is absolutely no doubt that they are attempting to find life, and when does one do that? ... only when there's an inkling."

"They are only attempting, mind you," reminded Athenia. "They have yet to find if life exists... maybe someday they'll succeed... but not now. I am in no way undermining their intellect or for that matter, their powers of observation; only stressing the importance of our intervention. Besides, think of the effect this inanimate microchip will have. Human beings are mortal; made of flesh and blood... they are sure to die, and their corporal body is sure to decay but not this insentient metallic microchip. It will continue to emit its dangerous rays for a long long time."

"How devastating!" Phoebia exclaimed in horror.

"And how devious!" added Athenia. "We need to be quicksilver to countermand their incursion. Having grandiose plans and good intentions are not enough; it's equally important to find a solution within the quagmire and execute it smartly and quickly. The moot questions are 'what' and 'how.'"

* * *

Phoebia and Athenia looked at each other. They were discussing a scheme to free the Earthlings from the malevolence of the Martians. They wanted to take no chances.

Athenia spoke. "We have already freed the afflicted human sample lying with us of the microchip and checked the other one for its presence. We now know for sure that the chip is no longer lodged in them. We have to move quickly to detect other similar samples scattered on Earth and work on them too. But, as I see it, that would not be enough."

"Then what else do we need to do?" enquired Phoebia.

"Destroy the very source from where it stems. We destroy the root, we destroy the very tree that grows from it," replied Athenia.

"And how do we do that?" asked Phoebia.

"We start by mapping out the exact location of the Martian premises which is a hubbub for such nefarious activities," answered Athenia.

"And then?" Phoebia prodded. She was always a little out of depth when discussing anything external to her theoretical wisdom.

"Once done, we'll have to destroy it," stated Athenia.

"Destroy it?" screamed Phoebia. "Are you hinting at destruction? Now it is my turn to say that the idea doesn't appeal to me. I find you extremely partial. Hurting the Earthlings is unacceptable, but when it comes to the Martians, 'destruction' is not necessarily a repugnant word."

"There you go," chanted Athenia. "Accusing me wrongly again. I have never found 'violence' appealing and I never will. But do we have a choice? Our intention is noble, and a noble intention justifies violence. There are no vested interests involved. The idea is neither to harm the Martians nor to obliterate their territory, but merely to subject their experimental sanctum to complete obliteration with a singular pointed attack. Something suspicious is happening there. What? I know not, but it sure is threatening and iniquitous for the future. Once their demonic territorial component has been exterminated, hopefully, there'll be peace in Space for a long time."

Phoebia regarded her superciliously, "But we need to have a Plan in place for that. Do we have one?"

"As of now, no,' replied Athenia. "But an appropriate stratagem can be worked out. All we need is an opulent conception to devise an attack that would efficaciously meet the exacting objectives. I speculated at the use of our most recent 'Long Range Fine tipped Missile' which we have only just finished assembling."

"Not a bad idea!" conceded Phoebia. "But it has never been tested on foreign soil."

"Now it will be," announced Athenia. "Our internal simulated assessments of the prototype have been a tremendous success, and it is due for its first trial in outer Space. Of course! The 'Red Planet' was never a part of our original Master Plan, but an opportunity is being offered on a platter, and it seems only wise that we make the best use of it."

"Are you sure the decision's judicious?" cross-examined Phoebia. "My allusion is not to the competence of the 'Missile' but to the paucity of time. We'll have to mobilize all the parameters within the limited framework that we have. Do you think that would be possible? Don't forget it's the first time that we will be testing it in alien soil. An outcome could tilt either ways - a victory or a failure. Nothing intermediary. A victory is always welcome, but a failure... are we prepared for that?"

"The mock demonstrations in our 'Work Space' have not given me any reason for vacillation," declared Athenia. "If I remember right, there was not a single *miss*. But yes, a few variables like *distance* and *force* may have to be recalculated if we have to ensure that the discharge is devoid of any flaw or slipup, but that's a minor deterrent. Remember, we are attuned to accomplish the impossible."

Phoebia found herself nodding metronomically to Athenia's assertions, but not before she had finished expressing her concerns, "I am completely sold on your earnestness, but my anxieties rest on the Martians. If it had been *yesterday,* I would have dismissed them as unimportant, but *today,* I'm wary of taking them for granted."

"What makes you so apprehensive? You were never like that," stated Athenia.

"The discovery of the human chip has made me so," explained Phoebia. "Their knowledge on the human body has shaken me so hard that you will have to forgive me for presuming that they have their sights firmly placed on every spatial pulse. We can no longer dismiss them lightly. Their ambitions are far beyond, perhaps, even our own Solar System. There exists every possibility of them keeping tabs on everything that's happening in 'Space' and that may include our Plan to outwit them."

Athenia stared at nothing and contemplated loudly, "Yes, it's a mortifying thought and we have no way of finding out the extent of their knowledge and probably never will, but as of now we are staring at a matter of the greatest urgency. It is contemporary and needs to be forestalled at the quickest. I want no time lost in adjudging their inventory of intelligence. If there's a risk.....it needs to be taken. No risks - no happy results."

There was a pause during which Phoebia reflected on what had just been said. Finally she relented, a little hesitantly, "There's wisdom in what you say, but the base...," she trailed leaving an unfinished sentence for Athenia to complete.

"That has already been decided upon," declared Athenia. "Even while you and I have been having this tête-à-tête, I have gotten to the extent of deciding the base from where we can launch the missile."

"Which one? There are two, remember," recapped Phoebia.

"The one that is a billion light years away from Mercury," replied Athenia.

"And the trajectory?" Phoebia queried.

Athenia contemplated for a while, before answering, "That needs to be reworked upon. In fact all the variables, as I reiterated earlier, will have to be reassessed and mathematically calculated, including the entry, the descent, the launch window and the landing technique; a time consuming drill I admit and a bit of a challenge too, but certainly not difficult. Our in depth know-how on distances, speeds and orbital paths between various Space bodies will see us through. We just have to be critically precise."

"So, when do we start cracking?" enquired a now geared up Phoebia.

"The sooner the better. Time is of the essence," said Athenia firmly.

"And these human bodies?" asked Phoebia.

"We'll deport them to where they belong," said Athenia. "Their disappearance may have flummoxed the Earthlings. Remember, it's not a case of one body vanishing into thin air now, but 2."

"One and Two?" Phoebia was nonplussed.

"The Martians chose to lift only one body at a time," explained Athenia. "We have lifted 2 from the same spot; one immediately after it was deposited by the Martians, and the other directly. The 2 make for an attention grabbing pair - they share the same parenthood and the same birth time with a difference of barely a few minutes. Down there, they are referred to as twins. It's an interesting inexplicable background for a potboiler."

"Fascinating!"

"Yes, spectacular is the word... Anyway, their human history is of little interest to us. What is, is the sinister artifice of the dubious Martians," declared Athenia with a firmness that indicated that she was in extreme hurry.

CHAPTER XV

Mr. Varma awoke to the sound of the sparrows that chirped on the sill of the window that had never been shut ever since the kids had evaporated into nothingness. It was a Sunday and his clock showed 7 a.m. The time didn't excite him anymore. In fact, nothing did. He had started feeling disconnected from his own life. Every day had seen him waking up uneasily or rather in a daze, making it more and more difficult to know if he was actually living a life, or merely stumbling around in some alienating nightmare that would give him yet another jolt. His intransigent fretfulness was beginning to wear on all those he interacted with.

Sundays had been 'fun' days, when the boys had been around. The mornings were usually spent watching reruns of old football matches and in the evenings, the family chose to visit a beach close by, to play the game with the rest of the neighbours who accompanied them. Once in a while, they visited the vicinal town, almost 20 miles away, to watch the local teams competing against each other. It made little sense getting nostalgic about those days, he thought to himself as he pulled up the blanket right up to his neck with the intention of going back to sleep.

He was just about to close his eyes when he thought he espied 2 figures in the garden. He sat upright, stood on his feet and walked across his bedroom towards the window that overlooked the garden. He rubbed his eyes vigorously to make sure that he was not hallucinating. He was convinced. Calling out to his wife exuberantly he asked her to follow him to the garden right away. There lay both his sons sleeping peacefully on the flower bed.

He tried awakening them, but they appeared to be in deep slumber. He checked their pulse. It was normal. He placed his ears against their chest. The heart beats were normal too. He couldn't believe his eyes. His sons were back and they were alive. It was a miracle. He shouted out to his neighbours and with their help carried them to their bedroom. He cautioned all women including his wife to distance themselves from the vision of the boys. He had heard in the media that boys who disappeared for a few days and returned safe, generally gave rise to a spate of female deaths, the moment they awakened. It was necessary to take precautions.

Knowing the intensity of hard work that the police had contributed in investigating his case, he dialled the local police station. The Inspector had just entered the cabin when he heard his phone ring. He lifted it rather sceptically. Of late, the cases were becoming more and more problematical, and he was unsure what this telephone call would entail. His astuteness coupled with his experience, which had in the past always assisted him in solving cases successfully, seemed to have deserted him. Picking up the receiver from the cradle he said, "Hello."

"Hello," said Mr. Varma excitedly in a voice that exhibited a puerile elation of discovery. "Inspector, I have some rejuvenating news for you. My boys have been found."

"What?" The Inspector literally shouted into the receiver both in shock and in excitement.

"My boys have been found," reiterated Mr. Varma.

"When? Where? How?" stuttered the Inspector.

"Why don't you come over? We can discuss it here, at my place."

* * *

In his dark room, Ayaz observed the pictures with an eagle eye. He was in an excellent mood. "My God!" he said to himself. "This is amazing. We may have struck on to something novel. It's important to get in touch with them immediately." He called up the Institute and requested for an audience with Rishi. "It's urgent. Something of prime concern," he told the operator who attended his call after he failed to get a response from Rishi's personal number.

Rishi cancelled his agenda and waited impatiently for Ayaz. Alex was there too. Not Diane. Pointing at the red colour splattered across the prints downloaded by Sridhar on that significant night, Ayaz outlined what he thought was a historic breakthrough, "Look at the colour. It is similar to the colour of the visual where the boys looked directly into the eyes of the robots and the human beings. I see a link - a glaring one, at that. What intrigues me though is the absence of any other element. If I'm not mistaken, you did mention the sighting of a uniquely designed aircraft, that night. But, there's nothing to indicate its presence in these pictures."

"Sridhar said the same," replied Rishi whose sense of reality had never ever been drowned in his schooled pragmatism. He was clear about specifics. He let his eyes wander to the photographs lying on his table. They were indelibly marked by Ayaz with specific dots highlighting the unknowable. As a professional, it was only logical that his conclusions be based on facts rather than fantasy, but when the pictures themselves were irrational, thinking rationally was evasive. He was in no mood to say anything, at least not now. He needed time to think over.

He turned to Ayaz, "Maybe, you could study these depictions, once again, in our studios here for a better understanding, and see if there's any variation in your inference. The facilities here are the best in the world."

It was a request rather than a command, and Ayaz complied. Acuities could vary in a changed environment.

But shifting his place of work did nothing to alter Ayaz's deduction. On the contrary, it only added to his discomfort of working, and he said so to Rishi, the next day.

"Pardon my saying this," Ayaz said looking at Rishi with an unspoken question that lingered in his mind, "but replacing the place of work has made no difference. My conclusions remain unchanged."

"Oh! I'm sorry if I have inconvenienced you," Rishi apologised. "There was no intention to undermine your infrastructure or your skill, but sometimes circumstances do alter insights. I thought being here, you could avail of our contributions, as and when desired. That was the only intention."

"I understand," responded Ayaz. "To come back to our subject of discussion, it's a pity we failed to capture the actual object. The only thing visible in these pictures is a trail of red smoke. It was this 'red' that I matched and analysed with those of the visuals I saw in your auditorium. The similarity is bewildering."

Ayaz hesitated and then proceeded, "It's an uncharacteristic colour."

"Uncharacteristic?" Rishi was nonplussed.

"Well! Let me be as explicit as I possibly can," clarified Ayaz. "There are colours and colours, and nobody has been able to count the number of colours that are found on Earth, but this shade beats them all. It certainly does not belong to our world. I say that with the experience of a man who's been in this field for donkey's years. Trust me or not."

"We trust you," said Rishi. "And the UFO? You mentioned that you saw nothing of it in the pictures!"

"No," replied Ayaz, emphatically. "Have a look yourself. It doesn't figure anywhere. There's just red and red everywhere, and that's strange. There should have been a hint of something, but there isn't. The images belong to the night when Sridhar sighted an UFO that apparently appeared from nowhere, don't they?"

"O Yes!" Rishi's mind ran quickly through the sequences of that night. "Even at such a distance away we couldn't help but notice that the strange object kept manoeuvring its sights over the boys. And all 3... Sorry! All 4 of us couldn't have been mistaken. And despite that, if it fails to get itself captured in the pictures then it is obvious that something idiosyncratic is happening. I know I sound absurd when I say that if logic is in no mood to assist us, then we may have to adopt nonconformity as a strong tool," he acquiesced.

"I think so too," prompted Ayaz. "A ghost, perhaps? Or may be an extraterrestrial."

"No. I'm making no such assertions. I'm only discounting the involvement of a human hand," he said with an aggressive panache that startled those around him.

Alex looked up sharply. "Goodness Rishi! Do you think it is judicious to base our judgement on such impulsive assertions? Remember we have yet to do an in depth research on the subject. It would be foolhardy to judge on the basis of only a single set of pictures that we got that night."

"I am making no judgement; merely telling you what the present set of pictures suggests. In the matching of the colours of the 2 different situations, there appears to be an interdependence," mulled Rishi.

"What 2 different situations are you referring to?"

"The sighting of the UFO and the blinking of the boys," said Rishi resolutely. "The absence of any tangible diagnosis, the absence of any convincing reason, and more so, the absence of any amorphous rationale is disconcerting. Add to it, the absence of any entity in the images, which is even more disconcerting. There appears to be an '*ABSENCE*' of everything here. Don't you think there's a need for reflection on the issue?"

"But this is so complex that we need time to reach a sane conclusion!" said Alex.

"Factually so!" reiterated Rishi. "It is so complex, that it does need time for a compos mentis inference. But you can't deny that it gives rise to a series of questions. Is it uncharted territory? If yes, then who do you think is behind this paradox? If the enemy is not among us, then where does it belong to? Outer Space? Another Galaxy? Another biome, yet to be discovered? If that's the case, we will have a tough time trying to wade the waters, and find the intent behind such a murderous action."

Rishi sounded effective, and his comment was succinct. No objections were raised as no auxiliary explanations were available.

"I agree," said Ayaz, solemnly, at last. "But then again, we cannot let vague assessments overshadow facts, so I would advise you to seek consensus from your research department. They are sure to have a collection of samples from outer Space or any other uncharted territory, as you'd prefer to call it. Why don't we just compare those with these? We may not necessarily solve the puzzle. We may never even get to find this shade in those pictures. But, at the least, we can let our lingering scepticism rest, and look for other solutions."

"You make it sound so simple; camouflaging a tricky situation with a sheen of simplicity. We'll work on that right away," said Rishi determinedly.

The dialogue was disrupted by the entry of Diane with an expression that was a hodgepodge of emotions not easy to untangle, "Heard the latest? The twins who had disappeared have been found."

"Found?" asked Rishi.

"Yes."

"Where?"

"In their own little garden, fast asleep," Diane said.

"Wasn't that the case with all those young boys who disappeared and were back a few days later?" asked Rishi

"It was," replied Diane. "But this time the news is causing a flutter for an inimitable reason. It is so sensational, that theories are being floated to give it a hard-nosed platform."

"The reason?" asked Rishi.

"People were getting so familiar with the tardy tale of the 'blinking eyes' that this new twist has hurtled the word 'enigma' to a different level altogether," proclaimed Diane. "It is now widely accepted that when an individual goes missing and returns after a few days, he's expected to awaken with blinking eyes, isn't it?"

"Precisely." Rishi looked curious.

"Well! They did return, but when they opened their eyes, they were not blinking at all," laughed Diane.

"What do you mean by not blinking? Have they returned with no eyelids?" asked a shocked Rishi.

"I mean they were blinking, but like us, and not like the others who were lost and then found. They are, if reports are to believed, normal, absolutely so. And the icing on the cake is that there are no reports of any female deaths like in the earlier cases."

Rishi was perplexed, "Really?"

"Really!" asserted Diane.

"The rapidity with which the events are unfolding is leaving me breathless," said a tired Rishi. "Everything is appearing so illegible, obscure and so unrelated, that I'm not sure if we will reach a satisfactory conclusion. Like the pigeon undergoing a catastrophic moult, this calls for a reconnoitring of our concepts, once again; a total overhaul of our theories, hypotheses, imaginations and presumptions. We will have to depute someone there to see if we can elicit some credible information from the boys that would fit into our sense of detail."

Diane shook her head disdainfully, "I am not sure that will help. There is something they have in common with the rest of their kind."

"And what is that?"

"Their unfathomable lapse," replied Diane. "They don't recall a thing. Their memory has been obliterated of the very period when they had been declared missing."

"Ah yes!" remarked Rishi loudly for the benefit of Diane. "The same happened with the others too, and that is precisely why we find ourselves running around in circles. If only they could remember! Isn't it baffling how the unexpected exigencies of life force you to realise that any form of control is only an illusion?"

Diane tilted her head, "Do you think a visit would be useful?"

"But of course! It would," replied Rishi. "And who better to do it than you."

"It's a risk I am not inclined to take," stated Diane firmly. "The world knows what happens to women after several interactions with such boys. We face death squarely on its face."

"But, I thought you said, there are no reported deaths this time," argued Rishi.

"Yes, you are right," conceded Diane. "There are no reported deaths, but you never know. They may react differently to someone like me. It's not easy to erase the trauma of facing these 'blinking' boys."

"Please, Diane, I stand guarantee for your safety," pleaded Rishi.

"Allow me some time to consider; only to consider, mind you. I want to emphasize that the response may not necessarily be in the affirmative," replied Diane.

"You have never disheartened anyone, have you?" said Rishi in an optimistic tone.

"Yes! The adventurous streak in me refuses to deny a challenge," acceded Diane.

Her sense of adventure peaked, and so did her vibrancy.

In the backdrop, the news reader of a major television network was issuing a warning alert for 'blinking boys' who held grave uncertainties about the future of the human race...

"*...If such a person is amidst you," continued the presenter," we request you to restore some of your idealistic values and convey the information to the Institute or your local government body. This is in the interest of the human race. The Institute believes that this may not necessarily be the handiwork of a human being. There is no strong evidence to back the theory at present but citizens are advised to...*"

...The rambling continued to include other mundane specifics like the location and the contact numbers of the Institute and the importance of sporting the innovative dark glasses that had been invented for the purpose.

"When did the Institute make such a claim?" asked Alex.

"I did," replied Rishi. The knowledge didn't frighten him – he was gradually becoming used to the unknown. "But we need to change our feed now, after what Diane has just said about boys returning with normal eyes. 'Tis a pity how even the most popular channels do not keep up with the latest."

"They are waiting for us to speak, for sheer authenticity," replied Diane. "I guess we'll have to undertake that trip."

***

A week later when the car pulled up outside the neat, compact bungalow of the Varmas, Diane and Alex had prepared themselves for a cannonade of commentariat. It was no wonder then that the serenity of the environs in which the house stood with its huge French windows looking out on to a small picturesque precinct, was a pleasant surprise. The owners seemed to have done a splendid job of keeping the Press at bay. The peculiarities of the astonishing happening were still fresh in the minds of the media. Under the guise of intellectual know-how, they continued to dissect it in a manner that made ordinary individuals feel that it was beyond the capacity of their average minds to grasp.

It was admirable how the family had taken pains to avoid making their residence a circus for all and sundry. Mr. Varma, a short, dapper man who stepped out to meet them had some difficulty in holding his composure. Nonetheless he displayed controlled courtesy in escorting them to the living room and discussing the preliminaries of the case over a cup of tea and some freshly baked bread.

In a voice as crisp as the unshorn desert hay, he looked directly into Diane's eyes and stated, "So, you were the one who had visited that boy Anil in school?"

"Yes," said Diane.

"We feel honoured to have you with us here. I was told that it was your keen observation that facilitated cognition for the innumerable female deaths, and associating it with the 'lost' boys who returned with this unfathomable 'blinking' oddity."

"Oh no!" replied Diane modestly speaking in a tone that she hoped was bland and confident. "Dr. Memon and Dr. Smith were the first to notice. I merely confirmed it, and the entire team at the Research Institute along with a few other independent medical practitioners worked hard to establish the connection. We have been working on this for quite some time. When it occasioned the first time we were, of course, taken aback, but passed it off as one of nature's freak incidents. But when it occurred quite a few times later, we became alert to a pattern that could not be dismissed lightly. It became clear that we had to find ways to curb it. We are optimistic we will succeed; at least our attempts are veering in a direction that will assist us in reaching a dependable closure."

"That's nice. It was kind of you to pay me a visit. I hope you find it worthwhile and productive."

"We hope so too," replied Diane.

"If you could give us a quick recap of what happened," intervened Alex. "We could then touch upon the finer details to get a fresh lead in the matter."

Recapitulating the events of the recent past, Mr. Varma went on to enumerate how his first son Karan went missing without a trace and how the local police along with the entire neighbourhood had lent their unqualified assistance in tracing him, but with no result. He had heard so much about teenage boys going missing for a few days and reappearing again with 'blinking' eyes that he had in his heart of hearts harboured a faint hope of seeing him after a few days. In fact, he and his wife had even gone out and bought those quaint spectacles which had been introduced in the market, for such an eventuality.

"Whatever optimistic thoughts I may have had about the shocking incident, insistent as they were, drowned totally on the aftermath of what happened immediately thereafter," Mr. Varma said resignedly.

He sipped some water from the glass that had been kept on the table and then proceeded, "It was a rude shock. As I said earlier we continued to cling on to some hope, no matter how modicum it may have been, and waited as patiently as we could. Certainly, Karan would appear; all the other boys did. But it was not to be. A few days later Kunal too went missing. He was playing in the garden, late in the evening, when I went to refill my pipe. That done, I returned to find myself standing at the window and looking out into an empty garden. It didn't occur to me that I would probably be reliving another horrendous replay. For an instant, I thought somebody around was playing a cruel joke. The old prickling seeped back. I stood transfixed for a moment until I motioned my wife to recheck. You can imagine our state of emotions. My wife lapsed into a stupor. Her sister had to be called in to take care of her. I was at my wits end, and the Inspector appeared to have greyed overnight. The entire situation was intriguing. I mean, how could this have happened and who could have done it?"

Humecting his dried lips, Mr. Varma continued, "We combed the entire area; the news was highlighted in the media as you are aware, and the Central Investigation was called in. Nothing came of it. There was no trace of him, not the slightest. Then mornings later around 7 a.m., I got up to have some water and just as I was about to get back to sleep, I saw 2 figures lying asleep on the flower bed. I was confused. I rubbed my eyes and looked out again. I was neither mistaken nor imagining. I called out to my wife, as also her sister both of whom were still in a state of shock. We rushed outside and what do you think we see? My two sons sound asleep, unhurt and peaceful. We tried waking them, but they didn't get up. We thought it wiser to call for the doctor right away."

"And then?" asked Alex.

"I was so sure I would see both the boys getting up with those funny 'blinking' eyes," claimed Mr. Varma, "that I advised both my wife and her sister to stay away from the scene. I was in no mood to lose any of the female members in the house. Of course! I had also kept those dark glasses ready as a precautionary measure."

"Gripping indeed!" said Diane.

"Yes," agreed Mr. Varma, "It does sound enthralling now, as I relate it to you but what I say is only a highly selective edited version of what I had actually experienced then. Imagine my trauma! My narration may sound pretentious, but to be honest, it is actually a watered-down version of my personal ordeal. All of which I say now is substantially accurate. Had I been a novelist, it would have done me much credit to pen down my sentiments, and when I say this I refer to my state of emotions that were running riot, at that time. But it's good to be impersonal, as I am today. Impassivity helps, especially when one is trying to relate something that people thing is a far cry from reality."

"I agree," Diane said, "and I empathize with you."

"Thanks," acknowledged Mr. Varma with a slight gentlemanly bow of his head and then continued. "I was both excited and dejected. Excited because I had found my sons, and dejected because I was unsure how long it would be before I would lose the female members of my family. Of course! As I had said earlier, the eye accessories had been kept ready to avoid such an eventuality, but the vagaries of life are unpredictable and one never knows when one could run out of luck. The boys continued to be asleep when the doctor came in. He contrived to examine them in his most mechanical fashion, camouflaging his excitement under a garb of questions that had no relevance to the situation. I remember him mentioning that he would give a qualified opinion only after the boys had awakened."

"It's all so fascinating. Then when did they get up?" asked Alex.

"Almost 12 hours later," replied Mr. Varma.

"What about the doctor?" Diane asked.

"He continued to linger. The more he lingered the more he enthused in the case. There has been such a hullabaloo over this weird phenomenon of young boys disappearing and returning with 'blinking eyes' that he did not wish to get away from the scene until well after the boys had got up."

"O my!" sighed Diane.

"Please continue," said Alex.

Mr. Varma continued, "When they finally did get up, they failed to understand what the fuss was all about. Only Kunal's reaction was noteworthy. He kept pestering us to tell us where we found Karan. Remember he had been very much around when Karan had got lost. Otherwise, the two of them reacted in very much the same manner. They felt they had just awakened after a long sleep and were greatly perturbed by what they thought was hosing them down with our presence, our eyes, our questions and what have you. The most startling thing was their freshness that came like a jolt in the blue. They behaved as if nothing untoward had happened, but I and so did a few others, noticed a celestial quality to their human ordinariness. The doctor and I looked askance at each other. In the exhilaration, we had ignored their eyes totally. When we realised this, we both looked at their eyes simultaneously; not once, not twice but thrice, but we found nothing strange. They were blinking as normally as all of us were."

Alex sat upright and so did Diane.

"Suddenly, I felt unsure of myself," resumed Mr. Varma, "for some strange reason I had an odd feeling that perhaps this time there could be something abnormal with some other organ of the body. I watched with restlessness the dissipating of my confidence which had briefly returned and requested the doctor to do a thorough check-up. "Nothing wrong; absolutely normal," he remarked with an assurance of a smug teacher trying to pat a student who had done his homework. I looked at the doctor and felt like an idiot when I asked him if I should bring my wife in, just to check. The doctor, a recognised expert in his field reasoned, "Females died, not during the first interaction but after 9 or more interactions subsequently," he said, a little irritatingly."

"Then?" prodded Diane.

"I was apprehensive. The doctor reassured me that nothing much would happen at the first meeting. The reaction would be limited and the risk acceptable. In the end all that she would do was to bend a little and feel a sharp piercing pain at the lower back."

"He's right," confirmed Diane.

"How do you know that?" asked Mr. Varma looking straight into her eyes.

"I experienced it," replied Diane.

"With that school boy Anil?" asked Mr. Varma.

"No, my meeting with him was the second. During that meeting, the pain at the lower back was even more acute than it was during the first."

Mr. Varma looked at her closely, "When was the first?"

"It was with a boy called Gattu who had undergone the same ordeal that Anil had, but in a different environment altogether."

"O really!" Mr. Varma said, barely able to contain his astonishment, "you must tell me more of this later."

"I will," said Diane. "But now, let's continue with what you have to say."

"O yes. Where was I? Ah! The doctor! He advised me to call my wife in. We did, but were unprepared for her uncharacteristically insane behaviour. She rushed towards the boys so impetuously that I had to literally tear her away from them to rein in her ardour and make her stand at a distance. The boys subjected her to an intense scrutiny as per the doctor's advice but there was nothing infelicitous. She exhibited none of the symptoms so vociferously discussed in the Press in a language that cloaked itself in intellectual legitimacy to something as simple as a stinging pain and a bend of the lower back. We were now convinced that they were not abnormal. The special eyeglasses were stowed away."

"Oh!" said Alex and Diane in unison.

"And there's something else I would like to add," continued Mr. Varma, looking directly at both of them, "My boys have undergone a transformation that is unbelievable."

Diane looked at him sharply. She found herself being swamped by a medley of emotions - anxiety, excitement, resignation, hope or all of it. She was unsure.

Mr. Varma continued with his chronicling. "Their notoriety which had always been the talk of this town, has now been superseded by a calm timbre. Everybody and anybody without an exception, has been commenting on their exemplary behaviour. This composure that they are exhibiting now, is so preternatural that I feel justified in saying that I am living a dream. I'm not sure how long it'll last."

Diane looked curious. "Can I meet them?"

"Sure. I'll call for them immediately."

The boys were well behaved and greeted both Diane and Alex as all well behaved boys do. Diane was struck by the fact that she felt no pain when she looked into their eyes. And yes, Mr. Varma was right, they didn't blink unnaturally either. Preliminaries over, the boys bid their byes to the guests and Mr. Varma continued the conversation like a punctilious host.

"Yes! It is indeed a reasonably straightforward case. There seems to be nothing more that we can add to what we have already heard in the media. I don't see why we should continue to stay on here anymore. Our early exit from here would be more constructive than our prolonged stay. We have done a lot of research on the subject, but now we may have to steer away from the obvious. We'll have to start swimming with the current flow of events which is so contrary to what we have seen in the past," said Diane at length.

"I agree with Diane," added Alex. "In the light of what has happened and what we have witnessed, we will have to start working on the basics again. You have been extremely generous with your time, but we should move now. It's important to restart at the earliest."

Mr. Varma demurred, "No, no. It is very crucial that you spend the night with us. There's something I wish to share with you; something that has been deliberately withheld from the media for fear of being continually hounded by them. The family is in no mood for interviews, opinions and incessant telephone calls."

Diane's curiosity was aroused. She exchanged glances with Alex. Had they stumbled upon some new clue? There was no harm in lingering if their extended stay would lead them to some clearer specifics on the theory they were working on.

"I gather, then," said Diane, "that it is something which has never been observed before."

"Frankly speaking, yes. At least, it has never been reported in the past. Technical jargon is not my cup of tea and I can only speak as a layman. I leave it to your discretion to decide. I may sound obstreperous, but until such time that I show you what needs to be shown, I can only request you to be comfortable. We have arranged 2 rooms for you. I advise you to get some rest. You may have to stay awake all night."

"Can't you share the particulars now?" asked Alex.

"The impact would be missing," stressed Mr. Varma.

"All right," replied Alex.

That night, Mr. Varma, after satisfying himself that the boys had fallen asleep, tiptoed into their room followed by Diane and Alex. The lights had been switched off and the room was mantled in complete darkness. The windows of the room were closed, and the curtains drawn. The only sound was the slight buzz of the AC.

"Just observe their heads," whispered Mr. Varma.

Both Diane and Alex looked keenly. For a while they stood mesmerized. There was a bluish colour in the area above the right ear of the boys. It looked luminous; almost as if a diamond was faintly sparkling.

"Oh my!" gasped Diane. She felt a quaint excitement which she subsided by taking long silent breaths to compel it into a self-controlled calm. Mr. Varma broke the spell in a conspiratorial tone which was a mere hush, "This is a first isn't it? Will this in any way establish a diversified route of thought?"

Diane inclined her head in assent and Alex muted a scream. Reluctantly all 3 of them walked silently out of the room. Diane was wordless, and so was Alex. Back in the hall, conversations were withheld for what seemed to be an eternity.

Mr. Varma was the first to speak, "Enthralling, isn't it? I noticed it when I entered their room on the night of their return, to switch on the bedside lamp. At first, I thought that my observation was merely a take off for my flight of imagination. I went nearer to have a closer look, and then it dawned on me that this was for real. I toyed with the idea of imparting the news byte to the media, but then decided against it. The next day I received a call from you. Being aware of your credentials for discretion and confidentiality, it seemed only appropriate that I discuss it with you before opening up to others. It's novel and I can firmly say with confidence that it occurred while they went missing."

Diane looked at Mr. Varma affirmatively and turned to Alex, "This calls for a discussion with Rishi and a request to send Ayaz over. He would freeze these wraithlike images into a series of photographs and throw some light on the subject, just the way he did about that red hue."

Alex agreed.

Ayaz flew in the next day, all excited. He had whistled appreciatively when Rishi had handed over a special camera for the shoot. It was a real beauty, not available in the market and of the kind he had never seen before. All he had to do was just point and shoot. He wondered where they got such stuff from. The excitement of working with these young scientists was indeed enthralling. Were they scientists or were they research professionals or something else? He was unsure of the right terminology to describe them but who cared? And now as he attempted to identify the outlandish luminous bluish tinge, it occurred to him that everything had been dreamlike ever since he had got associated with them.

The next morning Diane, Alex and Ayaz prepared to leave. They shook hands with the host, thanking him profusely for his warm hospitality.

"I hope you keep this to yourself," implored Mr. Varma failing to cloak his edginess. "Absolute secrecy, is what I ask of you. I honestly cannot entrust it to the 'Press.' It makes the boys nervous to be surrounded by these dapper neurotics who put on a pretence of knowing it all. It may garner unprecedented TRPs for the channels; but for the boys it would be sheer hell, because they still fail to understand what the fuss is all about. It's astonishing! How that particular 'missing' phase has bilked their memory altogether!"

"Don't worry," said Diane. "We respect your privacy. It will stay with us exclusively, only to aid us in meeting our intent."

Ayaz reassured Mr. Varma further by giving him a solicitous hug and promised him that they would do as they were told. "We don't know if we will ever meet again. Ours is an unpredictable life that has no reliable pattern, but your warm cordiality has touched our hearts. Your secret is ours to keep; only to be used with utmost discretion."

Having said their formal good-byes, Diane, Alex and Ayaz headed towards the local airport to catch their next flight, each one complacent in his own personal experience that was not likely to be forgotten for a long long time. Diane was glad she was returning with no pain in her lower back. She hoped that her painless back was there to stay. Was she mistaken in thinking that she felt healed when she looked into the boys' eyes? She was sure that she had divined a strange but friendly 'feel' - a 'feel of comradeship.' There was also an ever so faint smell of a flame that had just died. Was someone trying to give her a message? Well! Maybe, maybe not. She was uncertain, unwilling to share it with anyone. At this moment of time, all she wanted was to go back home and rest. Perhaps, she was tired... a little too tired.

As for Ayaz, he was mighty pleased with himself. He was gratified to note that a moment had presented itself - a moment that questioned his creative self-esteem. He couldn't wait to return home and have another look at the pictures that he had clicked. And Alex... he was wondering at the strange calmness that pervaded his entire being.

* * *

It was a little after 7, the next morning, when Diane stretched herself and looked out of the window. She felt well-rested. She twisted herself hard to check for the slight stab on the lower back, mnemonic of her meetings with Gattu and Anil, and found that it had gone. She stretched forward, backwards, sideways and all the ways that she could possibly do, but there was no pain. The temporary relief she had felt on leaving the Varmas' residence, was not temporary after all. She had been correct in inferring the healing effect she had encountered when looking into the boys' eyes. In fact, she thought she had felt a slight fading of the pain even then but had attributed it to her imagination and to the weirdly traumatic twists and turns of the day. Now she was sure. The boys did have a therapeutic effect on her. She had to share it with Rishi. The addendum would lead them on a different path, in their research.

Rishi was already in animated discussion with Ayaz when she stepped into his cabin. Ayaz had not slept all night. He had spent it developing the pictures that he had shot. He seemed extremely excited.

"Join us, and yes, bring your coffee along. Ayaz feels he has something exciting to show us. Bring Alex along too," Rishi hollered.

"Has he come to office? Isn't it customary for him to report late the following day every time he comes back from a fact-finding mission," Diane stated blandly.

"Not sure if he has come in yet, but I'm sure you should avoid making such statements in his presence. He may take offence to that indictment," admonished Rishi. "Anyway, if you do find him, ask him to join us; his combative questioning may offer a fresh facet to our disquisition."

"I will," Diane replied.

At the table, Ayaz compared the photographs taken at Mr. Varma's residence with the prints downloaded by Sridhar a few days ago.

"Have a close look at these colours," Ayaz said. The others looked at them studiously. The mini forum provided a useful opportunity for an exchange of ideas. "The visuals frozen by Sridhar," Ayaz explained, "and the stills which we saw in the laboratory have a red hue, but not the ones clicked at Mr. Varma's house. They exhibit a bluish tinge. And this blue colour, like the red, is totally alien to those that belong to Earth. Explicitly outlandish! Yes, explicitly outlandish!"

The silence that followed his thought-provoking observation was read by him as a confirmation of his assertion.

"I wish to emphasize, at the cost of repetition," he resumed "that all through my career, I have never come across such shades. I say it with the authority of an expert who has an undiluted sense of colour and an undisputed skill for segregating them. I hope my professional expertise is not in doubt and you accept my opinion not too lightly."

Photojournalistic courtesy obliged him to take a significant pause before starting from where he left.

"Apart from that, my surveillance has thrown up something very unusual that deserves a special mention. My approach to my job has always been no-nonsense, and I am perfectly positioned to say what I am about to state now. It may disturb you, but it's an input that may have some connotation to your research on the subject. While developing the negatives, those that showed evidence of a reddish tinge had an aggressive influence on me. I was so easily irritable and tired that I was compelled to take brief periods of rest as my eyes were being subjected to too much stress; not so, while working with the snap shots that were taken at the residence of Mr. Varma. Their influence was calming. I felt at peace while developing them."

Diane sat at the edge of the seat.

"Did you? This has certainly not caught me by surprise and when I say 'this' I refer to the 'influence'" said Diane pursing her lips before adding, "I was disinclined to share this, but now I think I should. It would probably contribute significantly to our subject."

"Yes, you should. Everything that appears insignificant assumes significance in the long run. Remember your grouse of a backache when you returned from Likwi," said Rishi. "Little did we think it would lead us this far. So, go ahead."

"I sensed a palliative effect when I looked into the boys' eyes while being introduced to them," said Diane, "and then that night when we walked into their bedroom to observe the faint bluish glow that appeared a little above their right ears, I felt healed. The back pain had gone. I presumed it was temporary because the euphoric confusion that followed left me extremely tired. I was in no mood for any sort of coherent contemplation and refrained from speaking much to anybody out there, not even Alex."

Diane paused significantly before proceeding, "All I wanted was to rest. And rest I did, when we returned. This morning when I got out of bed, I realised that the slight pain at the lower back had gone. It was not a transitory phase at all. Disbelief! That was the sentiment that raced through my mind. I turned right and then left; I bent forward to touch my feet and then backwards and lo behold! It wasn't there. I had been miraculously relieved of my abominable lower back pang."

"Interesting!" said Alex. "How I wish you had exchanged notes with me then. I too intuited tranquillity when I looked into their eyes; it left me puzzled and isolated. I attributed it to the weather. I should have shared this with you at that time. We could have spent a longer time with those boys instead of breaking our impromptu meetings into miniscule fragments."

"I wish I had dropped a hint, but at that time it did seem inconsequential. Not to worry though; we can always plan another more productive trip in the next few days with a more defined itinerary," said Diane.

She then turned towards Ayaz, "What about you?"

"I have nothing different to say. Forget the individuals; even the pictures have a tranquilizing influence," replied Ayaz.

"One more thing, Ayaz," continued Diane, "I guess we can rely on your discretion to confine this to the secrecy it deserves until we have identified the why, where, what and the how of this crusade. Let's keep our promise and not divulge a word of this to the world, please!"

"It's a promise I made to Mr. Varma and it's a promise I make to all of you," said Ayaz solemnly. Being a gentleman, highly talented and chivalrous to the point of ethical opacity, it suddenly prevailed upon him that he was now a part of a project that was a dark secret, to be closeted from the rest of the world. In his submerged simmering zest, his clarity of thought made him appreciate that his role was obvious in spiralling the youngsters to a triumphant end.

Rishi hearkened to the dialogue. What was bandied wasn't what happened in real life. It was unnatural. But one could not deny that it had happened, and one could not deny that something had to be done.

"This is getting to be more and more mystifying," said Rishi "Juxtapose the 2 situations. The first instance of lost boys who conclusively return after a few days to manifest their 'blinking' eyes on unsuspecting females. Result–death! The second instance of a situation where boys disappear for a few days and return with normal eyes. Result–Healing! Both instances are superficially abstruse but with an amorphously underlying complication. One destroys, the other heals. I cannot quite fix the resonance. But they have one thing in common. The disappearances are as unpredictable as the appearances and are under laced with a suddenness that indicates a dark possibility of subterfuge."

"You're right," avowed Alex.

CHAPTER XVI

Pontus returned to the basics. He was aware that the Earthlings were progressing in leaps and bounds, but he had not expected them to come up so quickly with an answer to the brainteaser that he had so painstakingly developed. The dark glasses were an interim measure to tide over the situation, and he had said so to Maxus in his last conversation with him. But it had succeeded in diminishing the effect of their latest invention. The thought disturbed him. His innovation had been challenged. He reworked the codes and looked at the chip through the 'Martian Magnifier,' and then at his team who were involved in its redesigning, "How long do you think before we are ready?" he asked.

"Shouldn't take too long," was the reply he got.

Maxus entered just then. "How's it going?" he queried.

"It's all set. Certainly not later than another Martian week's time, if our assumptions are correct." replied Pontus whose greatest professional asset was his phenomenal quickness in modifying an existing broad spectrum matrix to suit an orientation, as per its requirement.

"I thought it would be sooner," deposed Maxus. "You did say there was just a minor change that had to be integrated, but of course, you are the architect and you would know best."

"That was the idea," declared Pontus, "but as with most of my designs, there has been a last-minute modification. The present integration requires more than one module and to this end, specific quanta of mathematical chips have to be grossed up without hampering the fundamental schema. It will take some time to build in a sufficient variant to get the satisfactory outcome. The final design will provoke quicker results."

"Excellent!" Maxus knew that they were far too advanced. The Martian territory worked differently. No alien object in Space, whether a planet or any other, could bring their inexorable diligence to a halt. He pondered on the quantum theory of the state of an electron, according to which, farther an electron from its nucleus, more its energy. Is that why they were more dynamic than the Earthlings? But then what would explain the irritable force of the Venusians who certainly were at a higher rung of the 'ladder of advancement' as compared to the Earthlings? Did that mean Jupiter and Saturn were considerably superior? He made a mental note to make a study on this thesis, once their conquest of Earth was complete.

Pontus broke his wool-gathering with his monotone, "Just wait and watch. The redesigned chip will be far more effective and suave than its original. If it succeeds, and I'm sure it will, the Martians will be the protagonists of Space sooner than the prophesied deadline. The pugnacious Venusians would be obliged to discontinue their nosey parking."

"Don't count your chickens before they are hatched. Isn't that what the Earthlings would say?" said Maxus.

"No. I am not. I mean I'm not counting my chickens before they are hatched. That largely applies to the Earthlings who are rash in their behaviour, and short-sighted in their vision," asserted Pontus.

Maxus said nothing.

A Martian week later, Maxus found himself seated for the second time watching what was referred to as a rehearsal before the big 'Act.' As in the earlier case, a male robot walked into an enclosure but unlike in the previous occasion, he faced not 4 but several male and female robots.

"Multiple robots?" Maxus quizzed.

"Yes," Pontus replied. "We should have chosen to employ this earlier."

"Then why didn't we?" demanded Maxus. "You leave me befuddled. If the data base already consisted of the present codes in our array of formulae, and you were fully aware that we would get superior results by employing this equation, what deterred you from using it?"

"It's a case of the evasion of the conventional," deposed Pontus. "The choice to opt for the other, over this, was deliberate. I admit both my arithmetic and my estimation of the Earthlings went awry. But with the calamity that's staring right into our face today, we can no longer let everything rest on chance."

"Much time has been wasted because of our indecisiveness," derided Maxus.

"Not indecisiveness. Only error. It is crucial to note that trial and correction is central to our experimentation. It's not just focal but acceptable too," said Pontus to Maxus with an easy assurance of a student whose first brush with the unidentified was an essential step to success. "My presumption that an understated action would keep the dubious Venusians at bay, prompted me to act in an unstructured style. Besides, I had underrated the reaction of the Earthlings."

"Are you referring to those dark glasses, or the grouping of those 3 individuals in one spot?" asked Maxus in an indefinable tone.

"Here the reference is more in context to the invention of the eyewear," countered Pontus. "Much as our intelligence tunes out programmes to make other worlds appear obsolete, we cannot overlook the inescapable irony that our performance may not always be flawless. In such cases, the failure lies not in our incapability, but in our ignorance that other extraterrestrial races in Space are also speeding on the path of spatial-cyber sharpness."

"What prompts you to say that?" spurred Maxus.

"The Earthlings for instance!" answered a sobered Pontus. "It just escaped my deliberation that they would be smart enough to do some research, and invent those weird glasses with an inherent ability to deflect the chip rays from reaching the intended target. Why did I fail to amalgamate this while plugging in the concept? Of course! They have no knowledge on the chip, but their intellect did prompt them to build an obstacle to shield the victim from the offensive glance of the attacker. Let's give it to them. They are far more cerebral than we think they are."

Maxus nodded in understanding, "That piece of information should be treated as useful and should be borne in mind while carrying out the new strategy. We do not want to be thwarted by them again. Then it'll be a case of 2 against one."

"Two against one?"

"Venus and Earth against Mars."

Pontus was jolted to a rude shock. His involvement with the revampment of the chip had retroceded these planets to the background. While at work, he didn't care where the combat zone was pitched; his only interest lay in building a legacy within a given deadline, with a margin of flexibility to assimilate changes, if necessary. Now that he had shaped the chip to what he thought was a perfect ammunition for the war ahead, he shook himself out of the sopor.

So engrossed had he been in his task that these 2 planets, his prime combatants, had, momentarily receded to the background. But he was equipped with a powerful inbuilt mechanism which by default, never failed to engineer allusive resonances of the aliens in his travails. It steered him to a position of competitive mastery, most of the times.

"Shall we go ahead with the technical try-out?" was his only response.

Maxus nodded.

The multifariousness of the composition was neat, but misleading. Maxus was quick to observe the contrariety of the present presentation with the one that he had viewed earlier.

As in the earlier illustration the male robot strode in, but the dissimilitude lay in the number of robots that he had to cast his eyes upon. There was just one lone female robot in his field of vision. Behind her were positioned several imitations, both male and female. When the male robot set his eyes on the principal female robot she dilated her nostrils 3 times in quick succession, each time releasing moisture in the form of droplets. Those behind, placed their forefingers on their right nostrils.

Maxus straightened up. The chip had apparently been sorted and resorted through its original bits of encrypted magnetic information to transform its configuration considerably.

Pontus proffered an explanation, "For all appearances sake, the male robot seems to have activated only the female. But that is not so. If you had been keen enough, you would have noticed that the particles released from her nostrils evaded the atmosphere, and found their way to the secondary robots placed behind the principal one. The moment these particles reached the peripheral robot, some of them reacted with a touch of their forefingers on their right nostrils. This is an indication that the released particles have now been implanted in their bodies."

After a brief, perceptible pause he continued, "When the demonstration is carried out the second time, I would request you to observe, not the principal robot but the ones behind."

"Alright," acceded Maxus.

The second exposition saw the male model walking in, again, and looking piercingly into the eyes of the female. As in the preliminary revelation, the female involuntarily moved her nostrils in reaction. Maxus noted that he had been erroneous in thinking that all the secondary robots responded with a touch on their noses. Not so; only a few did and those that did were males, while the others merely flickered their eyelids. Another eccentricity of the recipients was their involuntary raising of the left arm and then letting it rest gently behind their back, bang on the spot where a throb was supposedly felt. The blink of the eye and the raise of the arm were simultaneous. It was a reaction common to all.

"The involution is alluring but the logic is beyond my comprehension. A simplification would help in relieving me of my ignorance, if I may be permitted to use that word. You may be conversant with the empiric finesse of Earth and its inhabitants, but my knowledge on the subject is hardly noteworthy," said Maxus, turning towards Pontus.

"I quite understand," submitted Pontus. "Despite its apparent intricacy, I've tried to come up with something that is relatively straightforward. The reaction of the main female robot that you witnessed is called 'sneezing' on Earth. It is an involuntary way of removing irritants from the body when contracted with a sinus problem. This time, the microchip has been fashioned to create multiple carriers through a solo action."

"Multiple carriers?" Maxus appeared puzzled.

"Yes," affirmed Pontus. "Let me break it down in the most simplistic way that I can. As in the earlier case, the introductory carrier will entrench the imprint of the specific chip in the victim through a series of Mars-electronic connections. The transfer will happen through a visual encounter."

"This part of the programme is familiar," said Maxus.

"That's because it remains unchanged," replied Pontus. "The modification sets in, during the second half. Unlike the last, the visual encounter, in this case, will stimulate a sneeze. The female will feel a pain in the back. Once entrenched, the imprint will transform itself from a latent to an operative one. Just as in the earlier programme, multiple interactions with the carrier will cause her death."

"This too is familiar," stated Maxus'

"Yes," agreed Pontus, "But this time she has been given an added responsibility. She will not just die but infect those around her with her seemingly harmless action of sneezing."

"Too obfuscated for my understanding," said an apparently confused Maxus.

"I disagree!" negated Pontus.

"Is it our latest innovation?" questioned Maxus.

"No, it isn't." Pontus proceeded to explain. "The codes in their primary form have always existed in our data bank. We have only juggled with the permutations and combinations. The complexity lies not in the reaction of the preliminary female target, but the concomitant effect it will have on all those present in her vicinity."

Maxus desisted from commenting. Instead, he sought a more elaborate exegesis from Pontus.

Pontus elaborated, "When humans sneeze, they spread aerosol droplets among the rest of their brethren. But this particular victim will not produce aerosol droplets. Instead she will spread miniature versions of the microchip, configured with 2 binders disguised as '*bio-bot viruses*.'"

"Two binders?" questioned Maxus.

"Yes," replied Pontus, "one for the female and the other for the male."

"A twist? Interesting!" prodded Maxus.

"In the earlier case, the rays bounced on sensing an 'XY' gene. The males were protected and had no role to play. Not so here! When the 'bio-bot virus' senses the 'XY' gene of a human male, it binds itself on to it and lies dormant until it sets its eyes on an 'XX' gene. It then gets activated and behaves just like the primary carrier, but to a lesser degree."

"Ah!" exclaimed Maxus. "An antithesis! The male continues to be active despite the temporary dormancy of the imprint."

"Perfectly," said a satisfied Pontus. "But not all males; only those who receive the virus."

"Those?" asked a nonplussed Maxus.

"You've nailed it right," said Pontus. "Sneezing is an action that releases particles in the air, and there is absolutely no guarantee that all of them will find a home in a human being, but some of them surely will. Most of them may just be blown away, and get dissipated in the atmosphere. The bio-bot viruses will remain active for a period of 6 hours from the time of their release, after which they'll cease to be effective, and disintegrate totally. Live human cells are crucial for their survival."

"Genius! But not without its fallacies. What is the guarantee that all of them may not be lost to weather?" asked Maxus. "The human characteristic for immunity necessitates a multiple interface before it succumbs to a bug."

Pontus reflected, "Your contention cannot be wholly disregarded. But you seem to forget that humans live in communities perpetuating interaction with the same people most of the time. The act of 'sneezing' will surely infect quite a few people in the surrounding environment."

"I have my doubts," expressed Maxus.

"In the previous case too we were uncertain. There were misgivings about the possibility of interaction with the same individual several times, but females died, didn't they?" said Pontus. "The same applies here, but with a difference. Earlier there was just one carrier, this time there will be more. The males lodging the malevolent bio-bot virus in their armature will use it inadvertently to target females."

"Utterly confusing," said a perplexed Maxus. The code was not what bugged him. It was the enforcement. He wasn't sure if he approved of something that provoked such a reaction. However, undermining Pontus was out of reckoning because he was aware that Pontus could create prodigies.

Pontus tried to disentangle the web of confusion, with a mathematical explanation, "Let us assume that a given environment has 4 variables, A, B, C and D. Let us further assume that A is the male who carries the chip and D is a male who does not carry the chip. Let B and C be the females. When male A looks into the eyes of female B, she will sneeze indicating that she has taken delivery of the chip. This action of sneezing will have an exponential effect. Female C and male D who are in the same environment will also take delivery of the bio-bot viruses via the aerosol droplets expelled by B."

"And then?" prodded Maxus.

"Male A will always be the primary carrier," explained Pontus. "After this action of 'sneezing' Male D will also become a carrier, but a secondary one. The efficacy of the *'bio-bot virus'* in D will however, stand diminished. He will cause deaths only after twice as many interactions with the females, as compared to the primary carrier."

"And the females?" intervened Maxus. He was still trying to grasp the functioning of this chip. He hated complications.

"Female B will always be the primary respondent," responded Pontus. After the 'sneezing', female C will also become a respondent but a secondary one. C will die only after twice as many exposures to the male carriers, as compared to the primary respondent."

"Completely baffling!" Maxus reiterated, still seeming puzzled.

Pontus attempted a simpler description, "If for instance, the primary male carrier manages to cause the death of a female after 6 interactions, the secondary male carrier will be able to achieve it only after 12 such interactions and the tertiary male carriers will be able to do so only after 18. The same is the case with the female victims. If the primary female victim dies after 6 interactions with the carrier, the secondary female victim will die after 12 such interactions and the tertiary female victim after 18. With every successive round of interaction, the potency of this virus will wane."

"I don't see how this will serve our purpose," enquired Maxus.

"It will!" Pontus replied with conviction. "Note that, this time, there is no obvious telltale anomaly like the unnatural blinking to lead the Earthlings to believe that something is deviant. Even if they suspect something is abnormal with the sneezing of the female victim, they will quarantine her and her only! The male carriers will never be in their sensor of suspicion and will continue to carry on the job that they have been unknowingly assigned to do."

"I find it purely quantitative," reaffirmed Maxus.

"It's qualitative too," Pontus countered. "Years of playing for the highest stakes in Space have groomed our skills to the optimum. I plan to have another practical exposition in the next couple of days, when you will be able to comprehend the imbroglio better."

***

A few days later Maxus was convinced that if not anything, the new plan had one compelling aspect; it was enormously unilateral. The results would be quicker, more tangible and far more measurable than its predecessor. It seemed straightforward and uncomplicated despite its exponential attribute. It had conviction for sure.

"Yes!" Maxus said, "It does appear as if the results may have a multiplier effect but don't trivialize their perspicacity. Don't forget, they already have a precedent and may be quick to jump to inferences."

"That has not been doused from my memory," Pontus said. "Even through the repugnance of the partial failure of the last project, I take solace in the fact that the Earthlings are amateurs. That gives us a decisive edge. We have to push ahead with our plans without bothering about unwarranted, groundless circumstances."

"Yes, we have to push ahead," agreed Maxus, vociferously. "We can't just let go of our goal of acquiring Earth. We can, and always will evaluate the menace of a threat, if any, by these mortals as and when it does crop up. We not only effectuate at a far superior level, but are realistic too. If demolition of Earth had been the only agenda, it could have been done a long time ago; in just a matter of a few Martian milliseconds. But it's not the planet that we want to obliterate, but the inhabitants."

Maxus's tone was clinical and surprisingly devoid of aggression.

"Agreed!" Pontus said. "Earth, per se, should never be destroyed. We want it exactly as it is with its vast expanse of blue waters, abundant growth of verdure, the imposing majestic appearance of high mountains, the wonderful layer of interesting atmosphere, the varied biomes and the endless bounty of nature. When we talk of spoliation, our allusion is to the populace. This selective attack on just the inhabitants, and not the planet itself, is making the job arduous for us."

"Not to forget the Venusians," reminded Maxus gently, "the self-proclaimed guardians of Space. Their constant obfuscating intrusion has inconceivably added to the glitches. We can't downplay their dominant role in obviating our efforts. The reminiscence of their last explosion which we had barely escaped by the skin of our teeth (is that how the humans would say?) still lingers fresh in my recollection."

"Ah yes! That was irksome! Those magnetic transmitters fizzling into choking vortices of sound and space dust would have left us frazzled, had we not agreed to abandon our operation," reminisced Pontus.

"Their expansive generosity towards Earth, to put in the mildest terms, is gallingly irritable. Why the soft stand, I wonder?" questioned Maxus in a barely inaudible tone. The question was like an intonation. It always lingered in the back of his mind, and surfaced every time there was a natter.

"It's all about a perfect balance in Space as I had elucidated to you earlier. That is why the need for covertness. Fighting 2 enemies indiscernibly leaves us with fewer options. Discretion is the key in our battle." Pontus sounded jaded and irritated. It annoyed him that Maxus unfailingly introduced the word 'Venus' in all of their chinwags. He controlled himself from giving a tart rejoinder and instead made a conscious effort to pay attention to Maxus.

"And you think this 'sneezing' plan will do the trick?" grilled Maxus. "I'm not being cynical; it's just that I'm disappointed at the outcome of our last endeavour. All I'm stipulating is success in this pristine machination. We control success, we control Space."

Maxus paused and changed the direction of his speech, "Any new development on that missing sample? The one of a pair of twins?"

"Well!" said Pontus. "The developments are not exactly pleasing. The body has been traced to the same place from where it was picked. It's alive, but not throwing up any signal. That has put us in a jam. We'll have to repick the sample and study it. Two possibilities exist; either the chip has been pilfered with or it has been wiped clean. Whatever, we have reason to worry."

"Is that a technical prerequisite?" asked Maxus. "I'm referring to the picking of the sample."

"Yes," replied Pontus. "Until we resolve that, the introduction of our 'sneezing design' will assume a lesser connotation. We do not want our new baby to be hijacked, in a similar way. A resolution would assist us in taking additional precautionary and preventive measures that may have been accidently overlooked."

"You think the Earthlings are responsible?" queried Maxus.

"No," replied Pontus. "If such was the case, they would never have resorted to the use of that ludicrous eyewear. The likelihood of a third entity is a robust suspicion."

"A third entity?" expressed Maxus.

"You're right. A third element," asserted Pontus. He knew exactly what Maxus would say next and he was right.

"Venus?" Maxus suggested with a slight bend of his head.

Pontus grimaced, "They unarguably were and always will be our prime suspects."

"Flabbergasting! It's turning out to be more and more convoluted than I had ever imagined," stated Maxus.

"Undeniably!" responded Pontus. "It's upsetting because this will only delay our modified strategy, but it is better to be safe than sorry. The hand of this element has to be corroborated before we plunge forward with the newest modification, so that we can incorporate their interference for an even more foolproof application."

* * *

Maxus was unhappy. He had not prognosticated so many snags. They had put a temporary brake in the Plans and yet, he was not dispirited. Breakdowns were an integral part of a tempestuous journey where the destination was known, but not the path.

"When is our next trip to Earth?" he asked.

"Not until the turbulence has died down there," replied Pontus. "Speculation is rife on the constant appearance and disappearance of one of their own. We can't fuel it further by giving them more fodder in the form of indisputable exploits."

"That would be a long wait," said an impatient Maxus.

"Yes. Perhaps longer than we think," grimaced Pontus. "Do you remember the spacecraft that had been orbiting our planet to collect data on us, sometime back? We had temporarily fooled them into thinking that the Sun had come between Mars and Earth, thus making them suffer from a communication blackout. Unfortunately, it has resumed its activities and is sloshing information all around. It has begun transmitting data to Earth again."

"Oh no!" exclaimed Maxus.

"Oh yes!" exclaimed Pontus, in return.

"How much information do you think it may have imparted, to date?" asked Maxus curiously.

"I don't know," replied Pontus, "but my latest reports say that there are plans afoot on Earth, to employ genetically engineered algae, bacteria and plants to potentially thicken our atmosphere by growing photosynthesising plants, bacteria and algae. They intend developing cell-sustaining colonies out here."

"They are amusing," stated Maxus.

"I thought so too, until recently," countered Pontus. "But now there's a question mark, and a huge one at that. Don't forget there are billions of organisms existing on Earth; there is a possibility that a few of them may lend themselves successfully to a fact-finding programme of this sort. But of course! I may be wrong in my conjecture since most of what I construe is based on the material sent clandestinely by the 'Max Probe' that has been sent to Earth. There's nothing very strong to substantiate it."

"That surmises the rationale behind the multiple trips from their end," Maxus conceded. "If what you say is authentic then regretfully, I'd have to agree with you, despite my disagreement in the past. It does make our job difficult. Any idea if they are aware of our 'Max Probe?'"

"I think not," replied Pontus emphatically. "It does sound ironical but their diversions in Space have yet to include the sighting of our 'Probe' which has been cleverly camouflaged. Ordinary people sometimes notice it and draw attention to it, but are snubbed patronizingly by their far superior, soi-distant lettered brethren who dismiss their sightings as a mere illusion. They choose to condescendingly refer to their finding as an 'UFO.' Others refer to it as an alien. But none have bothered to endorse the assertions. Little do they know that it's the truth. Remember! We are aliens to them!"

"An 'UFO'?" Maxus asked curiously. "Now what's that?"

"Unidentified Flying Object."

"Oh I see. And no one bothers to find out more? Not even the self-styled powerful intellectuals who are busy duplicating our Planet?" Maxus posed the question not to get an answer but to merely extend the conversation.

"That's the interesting part," said Pontus. "They are so busy applying their powerful intellect elsewhere that they have little time left to indulge on a trifling thing like an 'UFO' sighted by ordinary people."

"Trifling?"

"Yes, trifling," pronounced Pontus. "The Earthlings are like that. There are no equals there. They have the rich, they have the learned, they have the intellectuals and they have the rulers, all of whom group together to form the 'powerful.' It's an incomprehensible kind of power by our Martian standards. It annihilates the inferior whose opinion, observations and conclusions are rejected haughtily. That puts us in a commanding position. We can continue with the collection of our data unnoticed, unobserved and without the least interference. The conquest of Earth would force a provident step in our conquest of 'Venus.'"

Maxus concurred. "The planet 'Earth' finds itself in an extraordinary position. Besides being blessed with a breathtaking beauty, it is strategically situated to afford a convenience of both exploration and attack of Venus. It is pivotal, therefore, to transform our conviction from 'try' to 'have,' where 'Earth' is concerned. After all, 'Earth' is closer to 'Venus' than we are."

"Agreed," said Pontus. "If only we had a panacea to annihilate just the inhabitants and not its panorama..... things would have been a lot less Daedalian. Despite the several perfections that crown the existence of Earth, it also subsists with a bag load of imperfections which may paradoxically benefit us in the long run. But regrettably we cannot wait that long; hence the 'on and off' effort to conquer it prematurely."

"Imperfections?" asked Maxus quizzically.

"The planet is great but confusing," explained Pontus. "Ironically the foremost, among a few others, are the inhabitants themselves. They take so much for granted. If you place the 'before' and 'after' depiction of Earth in juxtaposition and compare the composite of incongruities, you will be fascinated by the difference. The green and blue that once dominated the scene has been replaced by grey, brown and a murky blue. The humans have tampered with the generosity of nature to build a skyline that will very soon dissolve in its own destructive construction. The abundance, that was so inherent in that planet, is being superseded by an overwhelming scarcity which will devour itself."

"Not to forget their greed to snatch everything from nature and give nothing, in return, to her," added Maxus.

"Ah yes!" conceded Pontus, "they have shrunk it to nothing. The teeming millions, the declining forestation, the adulterated ozone layer, the irrational exploitation of fossil fuels, the dwindling oxygen levels, the nuclear reactors, the frenzied fight for survival on the exploits of the less privileged, are all glaring facets of a planet that's running its final race. I'm amused by the Venusians obduracy to fight a persistent battle to preserve a declining planet that will be better safeguarded by us."

"Each to its own, I guess," said Maxus. "They are so blinded by our aggression and ambition that they are missing out the larger picture. Or maybe, they do have an ulterior motive - to annexe it for their benefit."

"Not so," decreed Pontus. "The rationale behind our conquest is to redress the natural cornucopia that our Galaxy has so abundantly bestowed on them. The intent behind their safeguard comes down to a matter of survival of human beings with females outnumbering the males. Do you remember the very first time we tried to attack Earth?"

"Very clearly," declared Maxus. "It was a revolutionary moment for us. It was hard to accept that we were actually defeated by the Venusians, but it did make us argute. We became wise to a simple fact that merely having great ideas, intellectual flexibility and technical superiority are not sufficient to score a victory. It is equally important to anticipate the enemy's move and counter it with a reconstituted expedient that will take the enemy by surprise and bring down their defences. At that time, our intransigence had exacerbated the situation, forcing us into a humiliating defeat."

Pontus broke into his narration, "We learn from our enemies too, but that does not take away the fact that if it had not been for the Venusians, we would now have been ruling the roost. The past cannot be undone. All that can be done is to concentrate on the present and on the times to come. It's all that matters."

"Agreed," accepted Maxus. "Utmost discernment - nothing wasteful and nothing imprecise to ensure that the mechanics and the timing of our future plans are punctilious and devoid of hiccups. Our plate has enough at the moment to keep the wheels of time moving with activities; Earth needs to be conquered and Venus controlled."

"But right now we have to stop that spaceship from relaying data to Earth," added Pontus. "Even as you and I talk, it is busy doing what we do not want it to do. We have to obstruct and block its presence on priority. The Earthlings are no longer as unreceptive as they were in the primeval days. Their frequent trips to our domain irrefutably insinuate their embryonic desire to hoard power and influence in Space."

"Ah yes!" exclaimed Maxus. "That's a dangerous development. Under no circumstance should they get wind of anything remotely connected to our Planet. Their malevolent enterprise has to be skittled with an immediacy it deserves. Press the 'DUST' button, as a temporary measure."

"That has already been done," declared Pontus. "The dust storm will do the trick and force the Earthlings' spacelab into a power saving mode, plunging it into perpetual blackness. But the measure is temporary and temporary measures are just that - temporary. Such temporary measures are lame for a matter as severe as this. A long-term strategy to counter its presence permanently, has to be chalked out exigently."

"Then find a permanent solution," commanded Maxus.

"That's exactly what we are trying to do," said Pontus. "We want no aliens probing into Mars. Certainly not the Earthlings who live in a world where perceptions outrace facts. The blueprint has been prepared. I've already called for a meeting of the 'Intelligentsia.' We need your inputs to give it a finality. A permanent closure to these irritatingly distracting objects from the Earth that interfere needlessly with our existence is a priority."

"To the exclusion of all other problems?" asked Maxus.

"As a solution and a buffer to all other problems," said Pontus, with an air of finality.

***

CHAPTER XVII

Diane stirred the soup slowly with her spoon. She was having dinner with Steve and Ayaz. The discussion was an annulus. It pirouetted around the recent rigmarole, and both Diane and Ayaz were attempting to enforce a response from Steve who, they hoped, would optimize their present impasse with his uncharacteristic point of view. They had run out of solutions and were seeking fresh view points from individuals outside the arena of their work sphere, in an attempt to solve the catch 22 locus.

Steve was a sober, practical person, whose only dream was to raise a world of blissful eternity. As a rationalist, he was aware that such a world was a far-fetched dream, a virtual impossibility; but the idealist in him refused to admit it publicly. Keeping his aspirations to himself, he laboured persistently, never once losing sight of his vision because he liked what he was doing. His devotion to humanitarian causes was noteworthy. And now as he listened to both Diane and Ayaz on a subject that didn't really interest him, he wondered at the millions being spent on research merely to find out what was happening in 'Space.' It was a sheer waste of time and money.

There were way too many problems on Earth, just waiting to be resolved: scarcity of food, population explosion, pollution, environmental destruction, discrimination, floods, droughts, wars, conflicts and terrorism. The diversion of funds to Research Institutes to find the existence of life in outer Space or for that matter even the presence of extraterrestrial beings, was beyond the domain of his comprehension. What did it matter if there was life in Mars or not; or for that matter in Venus, or even Jupiter. It was more important to take care of lives inhabiting the World.

As he reflected, he was shaken into the present by Diane, "So what do you have to say on this? What does your hunch tell you?"

"My hunch?" Steve looked at her with disbelief.

Diane was perturbed. She knew that, as usual, he had only been pretending to listen to her. Steve looked at Diane intently. It was true that he had given scant consideration to her words but he had not entirely ignored them. Snatches of the conversation had played and replayed in his mind.

Phrasing his sentiments into text, he said, "If you people had to spend even half the currency and time in solving the problems of our planet, the World would have been a happier place today. What more does a human being need? Happiness and only happiness! Why create glitches by venturing into a space that is hardly of relevance to us?"

Ayaz and Diane looked at each other meaningfully. The tirade was all too familiar. So accustomed had they become to the monologue that it failed to have even the faintest impact on them. It had to be taken with a pinch of salt. All they wanted from him, right now, was an unprejudiced clue that would help them find a solution to the present Gordian knot. Everything else had to take a backseat including his ever so repetitive grouse against the supposed wasteful expenditure on 'Space Research.'

"I agree," said Diane in a reconciliatory tone. "I am aware that in your line of work, to be eleemosynary is an absolute attribute, but not so in ours. We constantly deal with thankless situations and besting those is often the best reward."

"And the remuneration?" asked Steve contemptuously. "You forget that you people are among the highest paid; and for what? ... To study the sky and the stars. That's a laugh!"

Diane was wordless for a minute. Steve had a way of making her guilty of her chosen profession, but as always she dismissed them.

"The world's a kaleidoscope of varied views, opinions and ideas," she rattled on, "and we are all entitled to have our independent dogmas and tenets. We, at the Institute, too feel we are working for the betterment of humanity; it's just that our expression of the same differs from yours. I agree, it's expensive, but we are contributing, aren't we?"

"In what way?" Steve asked contemptuously.

"In many ways," said Diane. "We are contributing to the eradication of so many diseases by finding out the source. We are attempting to make life easier for people by discovering newer fuels and better modes of transport. We are attempting to cleanse the environment of pollution. And now, if we are attempting to find out if there is another place to live in Space... I see nothing wrong with that. Will that not help our future? I think we are being fair and just."

Steve looked at her in total disbelief. "Justice? Don't feed me with all that hogwash. What justice?"

"See darling," Diane continued, "years of research may reward us with the discovery of a surrogate planet akin to Earth where we can comfortably transport half the population. Wouldn't that be a wonderful solution to most of the problems? With half the hoi polloi living there, there would be fewer people inhabiting each of the planets: more food to eat, lesser strife, more elbow room, fresher air and all else that's needed for their equanimity."

"And when would that be?" asked Steve.

"As of now, there is no answer. But it is a certified given that someday travel to outer Space and inhabitation there, will no longer be a pipedream," replied Diane resignedly.

Diane had been married to Steve for 5 years now. His untainted passion for altruistic subjects was precisely the aspect that she loved the most about him. True, he was cynical. She had agreed to his prerequisite to not having children of their own, because Steve staunchly believed that there were enough mouths to feed in the world. He felt that, if ever, the need for children did arise, they would adopt a few and give them a 'quality' life. Until then Diane would continue to work for the 'Research Institute' while he would relentlessly pursue his passion.

As she spent the next few moments filling him in on all the salient points, his appreciation of Diane's perspicaciousness swelled further. In the past he had persuaded her many times to apply it elsewhere. She had agreed to consider.

There was something special about her. Everything about her, right to the perfect styling of her clothing to the way she executed her details, were organized and near perfect. Diane shared his disdain for the current morally-illiterate generation whose pecuniary interests overshadowed compassion for humanity, but her individual response on the issue differed. While he plunged himself vigorously into activities that he presumed would change the world into a better place, she silently stood by boosting his efforts, refusing to be entwined in a pursuit that was not necessarily her calling.

'They also serve who only stand and wait,' so said John Milton, the famous poet. But Steve didn't want to play the role of an actor who served by only standing and waiting. He was there to act, and act he did. His achievements were many, but his restless nature refused to let him rest on his laurels. Through force of character and eloquence that became his natural elegance, he had been instrumental in devising strategies that reached productive conclusions. His work place was dominated by the lines of Robert Frost...

The woods are lovely dark and deep

But I have promises to keep,

And miles to go before I sleep

And miles to go before I sleep.

Five years of married life and 7 years of courtship was a long enough period, to know her well. She came from a respectable middle-class background. He had watched her transform from a rebellious tomboyish adolescent to a sensible, pleasant looking adult whose earlier mutinous approach to curing the world's injustices by blowing up the offenders, was superseded by a supercilious outlook, the instant she took up a career at the Institute. The job fitted her like a glove. The eclectic profession, he noted with pride, had transmuted her into a person who radiated an enviable bounciness. He admired her gumption. Very few women, he had come across, could think with such clarity as she did. Besides, she was not frivolous. She could hold her own in a man's world.

He was conscious of her talents and appreciated her all the more for it. What he disliked, though, was the wasteful use of it under a medley of misplaced ideals that were fastidiously orchestrated in a fallacious direction. If only she could divert all of her abilities in solving humanitarian issues, the world would be a better place to live in. Yes! If only! As of now, he just had to live with the thought that she enjoyed her work, and his best interests lay in pampering her. She would probably change as she got a little older. Most people did. Until then he was willing to wait.

He smiled briefly and tugged at his tie loosening it in an effort to lighten the atmosphere. In private, he admired his wife for her brainpower. No one could deny that he was extremely impatient where his job was concerned, but when it came to others he was astonishingly tolerant. He detested intrusion in another person's passion including that of his wife.

"Are you listening?" asked Diane tartly.

"I'm sorry," he said rather apologetically. "It's rather taxing for me to dwell on it. My thoughts keep straying on to the monetary part. I know I sound boringly repetitive when I keep saying that it's rather difficult and loathsome for me to see such huge funds going down the drain on a subject that, as you say, is likely to serve a futuristic vision when all of it could be diverted in serving the present."

"True, true," agreed Diane just to soften his mood. "Look here Steve, you are fully aware that both Ayaz and I have always been in tune with you on this. But right now, all we request of you is your excogitation, experience and insightfulness to tackle a problem for which there appears to be no solution."

'Ok," he said a little reluctantly, "What is it that you want to discuss?"

"The same subject that I had touched upon earlier when you returned from one of those Southeast Asian Nations," said Diane.

"Something about a 'blinking boy' and 'dying females?'" replied Steve, his voice dripping with sarcasm.

"You don't have to be so sarcastic. The matter has taken a serious turn now. You know that as well as I do. It's not just a case of a single boy any more. There appear to be more of them all over the world. We have come up with these unique pair of glasses that have solved the problem temporarily, but are still clueless as to the source of it. We suspect the involvement of an alien hand."

"An alien?" said Steve in total exasperation. "Now Diane, don't tell me you have begun reading those science fictions again? Or is this another way to disguise your incompetency? When you fail to find a solution, you blame it on the aliens."

"I am not joking. Ask Ayaz," she asserted.

"Yes," intervened Ayaz. "I was present at this recording of a live demonstration of 3 boys in action, and I found myself observing a faint weird red streak being directed towards individuals in their line of vision. What beguiled me was the lack of response from the males while the females reacted with a faint shriek and a slight bend. That's not all, and this may sound laughable, but what really struck me was the colour. It was peculiar, not something that I had ever seen in my entire career. I know enough of colours to evaluate them and at the cost of appearing preposterous I don't mind admitting that it was way-out."

"Later," Ayaz continued, "one of their colleagues, Sridhar, sighted an unidentified object that grazed past the sky so quickly that he distrusted his sighting. A few nights later, while he again kept skimming the sky to clear his mind of a web of self-doubts, he was joined by Rishi, Alex and Diane. They were on their way to the parking lot. This time, all of them were privy to an object, shaped almost like an 8-gon, veering towards the Earth. It kept encircling just above the area where these 'blinking' boys are presently stationed. When the lights were switched on, it flickered and disappeared out of view in lightning seconds. The webbed veins of red fulguration which striated the sky, in its wake, also fizzled dramatically in a 'wonder if it was there' second. Coincidence? Accident? One does not know. I'm referring to its disappearance when the area was flooded with lights."

Ayaz stopped and looked at Steve to see if he was paying attention. He was.

Ayaz continued, "Neither the prints nor the images downloaded by the CCD, revealed anything of significance. The short-filmed script was a disappointment. All it exhibited was a blurred hazy picture. What was worth noting, though, was the colour; the entire set of photographs appeared a peculiar 'red,' a 'red' that was similar to the colour of the streak that originated from the boys' eyes when they were compelled to look at the people in the experiment that I just mentioned."

Steve looked at both Ayaz and Diane carefully. He knew they were not lying and he doubted they'd play a prank on him. Of course! The yarn did sound ridiculous, but he knew the 2 of them well enough to insinuate any mischief on their part.

Diane noticed Steve's scowl but chose to overlook it as she picked up from where Ayaz stopped, "and a few days later when we visited this place where a pair of twins disappeared and reappeared dramatically, one for a longer time than the other, we noticed yet another thing which intimidated us. It robbed us of our sanity, momentarily. There was this faint blue light that glowed from the upper part of the right ear of the boys while they were asleep, the visibility of which was possible only in total darkness. Ayaz managed to incarcerate that too."

"Yes, I captured them on camera," intervened Ayaz. "A study of both the colours, i.e., the red and the blue in unison only served to strengthen my unwavering deduction that the colours were certainly not ours to claim."

"He's the best photographer we have in this part of the world," emphasized Diane, "and he should know. Ever since, we have been brainstorming to encapsulate every miniscule moment of our experience into some sort of a cohesive explanation that would make sense. No success. We are still at a loose end."

Steve's stance changed from the inattentive to the attentive. This was interesting. In spite of his condescension to a subject that he thought was frivolous, he was intrigued by the supposedly important role of an unknown entity that Diane referred to as alien.

Almost impulsively he remarked, "Mars!"

"Mars!" both Diane and Ayaz chorused loudly drawing the attention of other diners in the restaurant. What Steve suggested was inconceivable. Mars? Impossible? How could a man of his calibre suggest something so daft?

Diane glowered at him. "Don't fool around Steve. This is serious."

"I am serious," said Steve surprised at his own admission. "When you say alien and in the same breath speak about the preponderant colour red, the first thing that comes to my mind is the planet 'Mars.' I am a layman with neither the photographic experience to boast of, nor the extensive research background to prove why I mentioned 'Mars.' The answer was based on what we had learnt in school where we always referred to 'Mars' as the Red Planet. Take it or leave it."

"Mars!" repeated Diane emphatically, not attempting to conceal her irritability. "Are you telling me that the planet 'Mars' is interwoven in all this?"

"I didn't say that," explained Steve. "You asked me what's 'red' and 'alien' and the only answer that I could think of was 'Mars.' I am aware that only someone like me would be foolhardy enough to express such an incongruous opinion in the company of experts and the intelligentsia, but you asked for my opinion and I gave it."

"But Mars!" a riled Diane repeated for the second time, wondering if Steve was mocking at her or was assisting her in reaching a plausible solution. The answer was sensational defying all logic. The research done by the Institute showed no signs of life existing there. But despite its irrationality, she had to admit one thing.....Steve's opinion did initiate a novel dimension to the entire chapter.

"Come to think of it," said Steve brushing aside her contemptuous response, "this could be a cause to fight for. If females are being exterminated in this fashion, then we need to gear ourselves to find the reason behind it. With no females around how would procreation take place? This is menacing, and if it persistently dodges a solution it probably is an indication that it is the beginning of the end of the human race."

***

When Rishi entered his cabin the next morning, he found Diane seated there.

"Good morning," he greeted good-humouredly "and what brings you here, if I may ask?"

"A very good morning," she replied. She appeared to be more cheerful than usual. "Steve has just returned from one of those insane trips, you know."

"Has he? No wonder, you seem to be in a pleasant mood."

"I am," said Diane ecstatically. "His busy tours leave us with little time for each other. Hardly 15 to 18 weeks a year!"

"What's new on his front?" enquired Rishi.

"Nothing, but the usual," replied Diane. "As for me, it was good. Very good. We went out to dinner after a long time; a pleasant change after those lonely meals that I've been forcing myself to, in his absence."

"Where did you dine?"

"At the Chinese restaurant that opened last month. Ayaz accompanied us."

"He's in town too, is he?"

"Yes."

"And how was dinner?"

"Fantastic. You should visit it with your family, someday," suggested Diane.

"Done. I'll consider your suggestion," accepted Rishi.

"But more than the fare, it was the conversation that took the cake," specified Diane. "Both Ayaz and I waxed eloquent on the recent happenings regarding the 'blinking boys' and the weird red colour that could be associated only with them. We concluded our narration by asking him for his view from a layman's perspective, and what do you think he said?"

"No idea. One never knows with Steve," stated Rishi blandly.

"He opined that by summarising all the stray words, i.e., 'red, weird, unearthly, space' in perspective, and stringing them, he could instinctively think of just one word."

A big silence followed.

"And what is that?" asked Rishi.

Diane hesitated before answering, "Mars!"

The impact was dramatic.

"Mars?" Rishi smiled sardonically.

"Yes Mars!" chuckled Diane. "Both Ayaz and I had a hearty laugh because it was just like Steve to come up with something so insane and incredible. But later, after the initial confusion and the surprise of receiving an audaciously shocking answer, I reflected on the possibility. He could partly, just partly mind you, be right in his thinking."

Rishi pondered awhile. He thought the innuendo touched on the ludicrous. It was too amateurish an opinion, a tad too abstruse and pretentious in the cosmic view of things, and yet he couldn't dismiss it lightly. It seemed to reinforce Ayaz's comment. Besides, nobody else had anything calculable to offer. So, he pretended to give it a thought.

"We have nothing to lose by giving credence to his suggestion," he said, "but doesn't it seem very unlikely? I mean, how and why and in which way can 'Mars' be interlinked? And what could have interested that planet so much as to indulge in a despicable activity of this sort? Does not the very thought defy dependability?"

"I concur with you," conceded Diane. "It makes no sense at all. It is hardly a dependable base to derive our theories from, but when all else fails, instinct, more than reason, helps a person grope in the dark. We need to, sometimes, think out of the box. Playing in the same sand box all the time, only succeeds in making us run around in circles. Let's for a moment presume that 'Mars' is the clue. Then what next?"

Rishi found himself nodding in agreement, although conviction was far from his mind, "Exactly! What next? There's no life there, at least the latest data suggests that. Then say again, for the sake of taking our case forward, that there is life that is as yet waiting to be discovered by us, then how do we go about tracing their involvement? That is, if they are involved in any way."

"Do we have any samples from Mars?" queried Diane.

Rishi responded in the negative, "I don't know. Ayaz had suggested something similar earlier. He didn't specifically say 'Mars,' but he did refer to samples from outer Space, but that was a submission prompted by the idea of matching colours."

"Great minds think alike," said Diane. "If Ayaz made such an endorsement and so did Steve, there does appear to be a faint possibility that a solution may emerge from outer Space."

"Sounds logical!"

"Then why don't we do something on those lines. Start comparing. If not samples, do we at least have photographs? There must be some which the satellite may have sent back to us?" prompted Diane.

"Photographs?" contemplated Rishi. "Ah yes. I'll have to confirm that. You remember the spacecraft that was catapulted into Space to collect deets on Mars? It was blocked by the Sun, for a while, but has begun relaying data again. Perhaps we can get a few 'exposures' from the department. I only hope they accede to my request. You know how reluctant and possessive they are about all their dope. It'll take a lot of wheedling to convey that nothing is intended that will conceivably result in leaks."

"If your obsecration is counterbalanced with a placatory attitude, I'm sure they will comply with your request," replied Diane. "You're perfect at that and have always managed to have your way in the past. Besides, this time around, we have enough ammunition to build the momentum. Convince them of the importance of bringing the present predicament to a fruitful, logical closure."

"I'll try," said Rishi.

"Please," pleaded Diane. "Once we have the images, we can begin by examining them for any 'red' and analogizing it with the 'red' of the recondite streaks that seem to travel so quickly from the 'blinking boys' to their targets."

"The idea is appealing," Rishi replied but his body language was in defiance to his answer.

Diane's suggestion didn't quite inspirit him, but on reflection it occurred to him that there were simply too many 'ifs' and 'buts.' He asked himself as to why would someone want to just wipe out the female species? Was there an ulterior motive? Was there something that they should know and they didn't? The answer was an obvious 'yes.' The very first disappearance of that boy Gattu and his reappearance with his weird 'blinking' eyes may have been an accident, but what explanation could one offer for the consecutive vanishings and emergences of the other boys at alarming intervals? In spite of himself, he could swear that they were not mere coincidences. A rational answer eluded him.

As Rishi tried to piece together the inexplicable interludes that went far beyond the symmetry and synergy of normalcy, the enormity of the situation anaesthetized into an imperceptible fear. This was not merely a challenge to man's astuteness nor was it a game to be outplayed. In the context, looking for solutions out of the conventional square, was not a bad idea after all.

He, thus, reluctantly inveigled the relevant department to part with a few frozen images. Only 2 were given; on a precondition that the give and take would be shrouded in secrecy and used only for the sole purpose for which it was intended. After all, even the general public was unaware that they existed.

"I've been strictly instructed to confine these prints to our premises," he told Diane once he got the prints. "But how do I persuade Ayaz to work in our studios? He is bound to refuse."

"Why?" asked Diane, a little surprised at this admission.

"He had made that clear to me when I had requested him to assess the earlier photos here," replied Rishi. "He had reluctantly agreed then because I had ingrained into him, that the amenities at the Institute were a cut above. But when the quality of the pictures showed no variation, irrespective of where they were developed, he was very definite about his priorities. It was then that he had insisted on undertaking all future assignments at his work place only."

"Try convincing him again. You never know. He may comply with your request a second time too," encouraged Diane.

"No, I don't think so," emphasized Rishi. "His reasoning is simple. I remember him saying that no matter how sophisticated or state-of-the-art the facilities, at the end it boils down to just one thing - the ease and comfort of working. And he certainly finds his own studio a perfect setting for his photographic activities. The comfort and familiarity that his own workroom offers, eludes him in a foreign setting."

"Don't we all feel the same?" expressed Diane.

"We do," acceded Rishi. "He didn't seem too impressed with our set-up. The only thing that blew him away was the camera which I had lent him for a shoot at the Varmas' residence, He said it was one of a kind and wondered if he could have one like it. I said I'd consider."

"Then the only way to solve the predicament is to invite someone else?" offered Diane.

"That doesn't appeal to me either," stated Rishi. "Several days would be lost in explaining the existing state of affairs, and we can't afford that. And you know the new breed. They all work in abstract quantities with little sense of reality. And then, there's the matter of confidentiality. How much would we be able to trust them? The world has downgraded to a bunch of greedy mortals who would willingly sell themselves for a few extra material benefits. Ayaz is tried and trusted. Besides, he knows the thread."

"You do have a point there," acceded Diane.

Several arguments and counter arguments later and with a good deal of cajoling from Diane, the images were taken out of the premises under the paranoid custody of Rishi, despite his disinclination to do so.

"A perfect day," thought Rishi as Alex, Diane and he drove into the driveway of Ayaz's studio with Ayaz at the wheel. As much as he despised the thought of working out of his usual confines, he tried to settle himself into a persistent optimism that this mini trip would open doors to possibilities of a working solution.

To most people, Ayaz appeared flippant in conversation and attitude; but the few who knew him well, admired the adeptness with which he handled his profession. He worked with missionary seriousness and a carefulness that was critical, especially, in matters like this. His atelier was maintained like a 'clandestine' outfit due to the high profile nature of his job. His clients never had the opportunity to see each other, and his staff were generally kept out of sight and only introduced on such occasions that demanded their presence.

While Ayaz spent cloistered moments comparing the relevant images, Rishi, Alex and Diane waited, nerves itching, with strays of conversation that were mingled with a melange of emotions.

At the back of his mind Rishi hoped that this session would put finality to the word 'alien.' A known devil was better than an unknown angel. It was easier to work in au fait confines rather than mysterious virgin open spaces. Unfamiliarity brought with it a host of quandaries. But what if his hopes were dashed? What if truly the hand of an alien was involved? The very thought sent shivers down his spine.

He recalled the moment when he had given an offhand statement to the media of the possibility of the involvement of an anonymous hand which may not necessarily belong to a human. That had been a paralysing moment; he had said so for want of something better. He hoped his utterance of the past would not turn into a prophecy, today.

The reviews of the images heightened Ayaz's sense of whodunnit. "Good heavens!" he proclaimed with an almost inaudible gasp of breath, which failed not to draw the attention of the others.

"What is it?" chorused the 3 of them together.

"It matches. Steve assumed right. This is unbelievable! The 'red' does match with these images from Mars, but this could be pure coincidence. There's no point in stirring a hornet's nest until further investigation. But if you ask for my opinion, this could serve as a strong base to make fresh inroads into your existing investigation."

"Wow!" said Rishi. "There's a clue."

"What next?" Diane asked.

"Yes, what next?" echoed Alex. "We always seem to be asking what next and not moving beyond that."

"Let's get back to the workstation. This calls for a pressing round of discussion."

* * *

Later in the day, the 3 of them Rishi, Alex and Diane along with Ayaz foregathered in the mezzanine cabin. As they sipped their coffee, they realised that fighting an enemy whose existence was undetermined, was already a lost battle.

Rishi's mind laboured on. 'Aliens?' Do they really exist? Were they a threat? He thought that they were mere fodder for sci-fi movies? But now he wondered. Maybe there was some truth in the saying that truth is stranger than fiction. The incredible thing, he mentally conceded, was that despite its irrationality, it didn't appear incongruous. A few more questions, a little deeper reflection and some much-needed thoughtful retrospection might produce some useful answers.

"Although there apparently is a connection, I am wondering how to put 2 and 2 together to convince myself of a logical explanation. We will have to harmonize our hypothesis to a professionally perfervid and coarctate rationale, to include only the pertinent assumptions. Now, how do we do that?" Rishi said thinking out loud.

"Maybe we could start by reciting the chronology of events like we always do at the start of every colloquy. It is so critical to the nodus. It's like a mantra, a preface. So to recapitulate–first you have these boys who appear, sorry not merely appear but actually release an impossibly hard-to-discern red streak. Then you have this spaceship that seems to leave a trail of a similar colour. And now we have these pictures, of Mars, which display the analogous red colour. All of them display a comparable shade of the same colour that has certainly not been observed any time before on this Earth. Isn't it Ayaz?" asked Alex.

"Certainly, yes; that's something I'm doubly sure of and I say it with a big 'EMPHASIS,'" avowed Ayaz.

Rishi paused before voicing his opinion, "Ayaz's deduction is an eye-opener. Our strategies have, all along, been based on the theory that we are fighting a meddlesome pugnacious foe who may or may not be 'one' among us. Circumstances now remain changed. Our sense of complacency has somersaulted calling for an entire overhaul of our approach. Undoubtedly not an entirely happy one, but at the least it does proffer an idea to work upon."

"Yes," agreed Alex. "Until now we have been groping, struggling and hypothesizing, clueless to the origin of these devastations; but not anymore.

Not that the word 'Mars' is a solution, but it does present a new route for exploration... new clues, new deductions, new..." He stuttered before proceeding, "Well everything 'new'. To sum it all, it proffers a new idea to work upon, although not essentially a logical one."

"Yes! It does defy the empirical," admitted Diane, "but we can no longer remain unreceptive to a pioneering way of dealing with the mystery. But I'm still a little confused. We appear to have an explanation for the colour red but that, by itself, does not answer all our questions. There is still some scepticism about the shade of blue seen in Mr. Varma's house - how do we expound that?"

"Ah yes, that shade of blue!" iterated Ayaz, "There's not a slightest trace of that in all of these pictures." He paused as if to ponder. "But the dominance of one colour does not decry the presence of others. It would be unintelligible to associate only one shade with Mars. 'Red' may be their predominant tincture, among a host of other minor colours with lesser signification."

"I stand on agreement with that," said Rishi. "We generally associate the colours 'green' and 'blue' with Earth but that does not take away the fact that we have over a thousand colours to take pride in. Likewise, 'Red' may be their most prominent colour; another conclusive reason why we may have seen it several times while 'blue' was seen only once."

"Yes," agreed Alex, "in Mr. Varma's house. But it would be folly to ignore it in the light of the context that we had seen it and not just seen it, but also experienced its effect on all of us... a soothing, calming influence. And Ayaz concedes that the 2 colours had contrasting effects on him while developing the negatives. The 'red' made him aggressive and listless while the blue made him calm and collected. Am I right in assuming that, Ayaz?"

"Perfectly," said Ayaz. "But remember, colours on Earth also have similar effects on us. We associate blue with calmness, red with aggression, green with youth, yellow with creativity ..."

"Godammit! We only seem to be running around in circles," countered Rishi. "Once again, the unknown is causing despair. I'm in no mood to fight a losing battle. What do we do? Mars? How do we start collecting evidence on that?" Curling his hand into a fist, he banged it hard on to the glass table, hurting himself in the process. "Ouch!" he reacted pulling his hand off. "This does not hurt as much as the thought that someone out there is outstrategizing us. And the only lead we have is the word 'Mars.' Curse it! We are back to square one."

It took a while for them to gather their wits after the outburst and return to a semblance of equanimity so necessary for a clinical symposium.

"There's no point in letting ourselves drown in a mire of misconceptions. 'One thing at a time' would be a good way to start. At least the 'red' appears to have a strong link to strange phantasms on Earth," said Diane. "If you recall, Rishi, when we were with Sridhar that night, the spacecraft kept circling right above the location where the 3 boys are located. At that time, I presumed it was sheer coincidence. But now I am not so sure. Perhaps they felt a strange attraction to the boys because they sensed the presence of a shade like theirs."

"Or, perhaps, they are responsible for the streak..." said Alex thinking loudly for everyone's benefit. "Presumptions are aplenty; but answers... that's what we are seeking... the right answer that seems to evade us. If they indeed are responsible, then why did they do such a thing?"

"What thing?" Diane queried.

"Target females only!" replied Alex. "There's no logic in doing such a thing unless they need to prove a point. I know of races, on our own dear Earth, who show a distinct preference for male progeny. They indulge in condemnable practices like sex determination and female infanticides. The picture here is inconsistent. Why would anyone on Earth target women beyond their adolescence? We need them for our survival, don't we? Besides, they are now on par with us on every dimension—physically, mentally, intellectually, emotionally... in fact, I wouldn't fight shy to say that they are emerging to be far superior to the males."

"I had overlooked that. Even in areas that follow such dishonourable practices, the laws are in place. It's illegal and you can be penalised for it. So, I would totally and completely rule out the role of a human creature," gave in Diane.

"We can debate on that, at another moment of time," interrupted Rishi. "Let's get back to our subject of discussion; let us presume that an extraterrestrial is responsible. Under such circumstances our conclusion that the boys may have been hijacked only to be returned with the sole aim of eliminating females, raises a string of questions. Who are they? What is the reason for their antagonism for the female gender? Why did they kidnap boys who were very young, mediocre and in no way exceptional, at least by our standards? Why was the kidnapping so sensational? Why was the disappearance and reappearance staged with such abruptness, and without so much of a trace?" went on Rishi.

"Prejudices aside, my sixth sense tells me, despite fears expressed, that the answers lie beyond the gravity of Earth," reiterated Alex. "Pun intended," he added jocularly.

"Are you hazarding a guess or merely echoing Ayaz's thoughts?" asked Rishi.

"It's neither a guess nor an echo of Ayaz's observations. It's a conclusion. Who else can they be? Like Steve says they are probably from Mars; a mere hypothesis now and yet the only one for want of something saner. And even if we do presume that the Red Planet is responsible for the terror, then the next question that logically follows is - are there any living beings there? Only a living entity would be capable of such an act," said Alex.

"As far as I know; none," said Rishi. He was standing at an open window that overlooked the wide, open space of the Institute where they had met Sridhar a few nights ago, trying to conceal his hopelessness at the situation.

"There's been much interest shown on the subject," he stated unemotionally, "but none of the studies have revealed the existence of life, at least not so far. And with no living beings in existence there, chances of something like this happening are remote. My God! This is so exasperating. I guess there's a point beyond which analysis based on mere knowledge and experience can stretch. Steve's suggestion of 'Mars' may sound like a fantasy, but one cannot deny that the view has opened a new door to our investigation. Imagine basing our investigation on 'aliens' instead of prudence and rationale!"

He turned from the window and looked at Diane, "By the way what has been happening with the boys?"

"I'm glad you brought that up," replied Diane. "There's something of importance that I wanted to draw your attention to. Something about those eyeglasses. They appear to be losing their efficacy. The female attendant who has been deliberately retained by us, because of her gender, to do constant checks on them, uploaded a report last night. She complained of a slight stab in her lower back, and you know what this means - it means that they are losing their effect. We'll have to give a new pair to each of the boys. The eyeglasses seem to be capable of handling just so much, and nothing more. They have to be constantly replaced for them to be effective."

"That's bad news. The boys will have to be provided with not just a new pair, but an improved one at that," responded Rishi. He gave a slight shiver. "This is a dangerous state of affairs. Vicious and inhuman! The conclusion is loud and clear. The rays which emit from the boys' eyes have the power to gradually but surely erode every surface that they strike and that's a bad omen. If it is strong enough to erode a nonliving substance, think of what it could possibly do to a living being."

"The thought makes me completely ill at ease," shuddered Diane.

"It should," said Rishi. "During our demonstration, the rays may have boomeranged from the surface when the target was a male, but we can't deny that they did hit the target to affect it in an evil manner. Compare the two. Like the eyeglasses that have become less effective due to the erosion of the surface, so will the males. It is obvious now, that the aim of our 'enemy' whoever it is, is to destroy the human race and wipe it off totally from Earth."

"Cause for immense alarm! We have to be quick, very quick," said Alex.

"But what do we do? I feel so vulnerable," moaned Diane. Despite her forcefulness to influence Rishi into considering a Martian hand in the unfathomable events that defied a judicious vindication, she was uncertain.

***

CHAPTER XVIII

Sridhar had requested for a workstation that had a casement overlooking the vast expanse of the Institute. He was given one. He stopped stepping out thereafter. It had now become customary for him to look out of the window and check for anomalies, if any, and compare them with the images that were being downloaded.

This evening he did the usual. Just as he was surmising that the sky looked a brilliant blue studded with diamonds, it abruptly turned ashen without a warning. He could see no sky, no stars, nothing at all, not even the outline of the neighbouring building. Everything was cloaked by what appeared to be a strange greyish blue smoke. He peered, straining to penetrate through an invisible sheet of cover to catch a glimpse of an elusive clearer patch of sky and as he watched, the sky transformed itself into a kaleidoscope of swiftly moving overcast clouds eroding visibility to almost zero.

He had a premonition that a disaster was in the making. There were several times in the recent past that he had felt the same. He had even made careless jottings of the same in his yellow book and attempted to reconfirm it by letting himself into the open. This time though the opacity was severe. In retrospection he realised, a tad bit late, that such an occurrence should have been reported much earlier along with his observation, even at the cost of appearing fatuous. The hazy images that almost always coincided with the dates shared by Rishi had been discussed, but not the changes in the atmosphere on such days.

It was late; beyond half past 11. Not many of the white collar staff were likely to be around at this time, but his presumption that Rishi would still be, was right.

Rishi had just finished securing his drawer when the intercom buzzed. It was Sridhar. "Did you have a look outside?"

"Why? What's the matter? You sound flustered."

"The sky! It has turned an unnatural steely colour. Look out, and you will know what I mean."

Rishi rose and walked towards the window. When he had first met Sridhar, he had not been impressed. He had slotted him into the category of those modern-day business graduates who categorised solutions into neat packages. It was all black and white. Never grey. They were dismally low on emotional intelligence and field experience. But Sridhar had proved him wrong.

The scene outside put an abrupt end to his thoughts. It was dark. There was nothing unusual about that; it was expected to be dark at this time. It was just that the darkness had an eerie mantle and visibility had dwindled to an extent that he could not see even beyond 2 inches from himself. Only 2 hours earlier the weather report in a popular news channel had launched into a dissertation concerning temperatures and pressures. It was expected to be a pleasant night as per the forecast. What he noticed now, was to the contrary; luminosity was at a gloomy low and the slight breeze, that blew, plagued the night air with a psychic chill. He was stunned.

With the receiver of the intercom still in his hand, he fiddled with the long chord that connected it to the cradle and shouted, "Goodness Sridhar! What's happening? Is this happening only here or all over the world?'

"I don't know," replied Sridhar, "but one thing is for sure; the tenebrosity has never ever been of this intensity before."

"What do you mean by 'before'? Are you alluding that similar changes have been noticed by you in the past?" yelled Rishi.

"Yes, but never of this severity," Sridhar replied.

"Why was I not informed?"

"Shall discuss later. Right now, it's pertinent to observe what's happening outside."

And as they kept watching and talking, Rishi from the window of his cabin and Sridhar from his workstation, the sky became clearer. After about a good 20 minutes or so, it returned to its original blue, revealing a clarity so astonishing that the opacity which preceded it seemed like fiction. The stars winked impishly at them and everything returned to normal. But for the fact that both Rishi and Sridhar had witnessed it from 2 different points, each of them would have passed it off as a hallucination best forgotten.

Immediately after the night regained normalcy, Rishi stomped into Sridhar's work place with his mind plunging agitatedly into a whirlpool of questions. "Did we really see what we thought we saw?" he asked him in a fanatical tone.

Sridhar appeared out of sorts himself.

* * *

The Institute was in a bedlam of disbelieving commotion. Diane looked at both Alex and Rishi incredulously. "This is implausible! Truly outrageous! How could the boys have just disappeared altogether, traceless?"

"Strange! Very strange!" muttered Alex. "This is big trouble."

"Trouble? It's cataclysm! Trouble is what we have encountered in the past. This is graver and more consequential," ranted Diane.

"Hold on!" said Rishi. "In all this pandemonium it escaped my memory to share with you what occurred last night. I wonder if that has something to do with the disappearance of the boys, or if it's mere coincidence."

"What is it?" synchronized Diane and Alex.

"It was about 12 in the night. I was in the process of signing off for the day when Sridhar called on the intercom sounding all flustered. He beckoned me to look out of the fenetre which I had only just fastened. I was dumbfounded. The sky had turned fonce. It was arduous to see my own finger even from a distance of just 2 inches. This was accompanied by a distant rumbling threat of a tempest. The status quo remained unchanged for almost a thousand seconds and as we, that is Sridhar and I, kept peering through the obscurity, the sky slowly turned blue, once again. It was paranormal. Boy! Was I glad I was around at that time! Had Sridhar conveyed this to me the next morning that is, today, I would have thought that he had gone psychotic."

"Oh no!" moaned Diane. "The security guard conveyed the same to me. I was infuriated. I thought he had gone bonkers. I let go of a series of expletives. It was natural of me to do so, after the audacious magnitude of what he said. I need to apologise."

"What did he say?" asked Rishi.

"Exactly what you did, but a little differently. He was on the beat when the sky became unexpectedly dark, so dark that he had to stop himself on his tracks, as it was impossible for him to put even a step forward in the direction of his movement. This was accompanied by a continuous deep resonant sound. Strangely, even the night dogs didn't bark."

"And then?" prodded Alex.

"He was flabbergasted and wondered what to do next when, without a warning, the obscurity gave way for a clear blue sky. He looked at his watch. It was almost 12. It struck him that the intermezzo had lasted for barely 15 to 20 minutes, although it felt like eternity at that time. With a slight dismissal of his shoulders, he continued on his rounds with a sombre attitude, until he was forced to stop right on his tracks. He stood rooted to the spot with his mouth agape."

"What forced him to do that?" asked Alex unable to contain his emotions.

"The door," announced Diane.

"The door!" exclaimed Alex.

"Yes. The door leading to the boys' dormitory was open," related Diane. "The boys had evanesced. They ferreted for footprints. There were none. They looked around for telltale signs; none were found. There was no suggestion of a forced ingress. The door seemed to have been opened with some laser beam."

"What else?" Alex probed.

"A faint smell closely resembling the scent of a dying flame of an unplaceable bittersweet tang mantled the room. Nothing is missing, and nothing unfamiliar is left behind. Everything is intact," said Diane. "But for the boys. What do we do now?" She gestured frantically in a manner that only served to accentuate her desperateness.

"No clue, at all," said Rishi resignedly. "Do you recall the conversation we had when the twins disappeared? It was one initially, then 2, now 3 and in the near future, 4. The numbers of disappearing males will increase and we'll still be trying to figure out who is responsible. This is becoming increasingly threatening, and we are yet nowhere close to solving the riddle. Sridhar says there is no cause for worry. The boys will be back just as promptly as they had disappeared."

"The only optimist among us," said Diane, ironically.

"We had an interesting conversation last night during which he acquainted me with an unusual observation he had made during his study of the sky. It was an eye-opener. To believe it or not, is the question," said Rishi.

"What is it?"

"He briefly explained to me how he had, on occasions, detected a change in the colour of the sky and how they almost always quadrated with the disappearance and reappearance of those 'blinking' boys. This time around, though, he thought that the transformation was even more austere, and he attributes it to the fact that 3 of them disappeared at once. He's certain that it wouldn't be long before we witness another incongruous change in the pigmentation of the sky resulting in the reappearance of the boys."

"Nice fairy tale," said Diane.

"I thought so too," stated Rishi, "until I realised that he is more au courant with the vagaries of the atmosphere than we are. Besides, he is one of those serious guys who cares not for anything frothy. It would be unwise of us to take his grim assessment of the situation with a pinch of salt."

"I agree with you; the lad's a genius. Mark my word, one day he will be a 'figure' to be reckoned with," said Alex who had always had a soft corner for the young man. "For the moment, let's be stoical and see what happens next. This may sound illogical, but my certified opinion is that the red 8-gon shaped UFO was indubitably targeting those boys. The occurrence of this event only confirms my qualms, which brings us to the mother of all questions that has been a regular in all our talks; what do we do now?"

"Start with clues and guesswork like a detective pinning suspicion on everyone, every incident, every hour and every what have you. It's a pity our store of facts is frugal, which, ironically, is again based on unproven theories, diffuse whiffs and unreliable scenting. For now, we can only rely on suppositions. And yes Diane, not a word of this to the media. Please inform everybody that the ongoings of the past few hours must stay within the premises. They have to be incommunicado. Ethics, prevarication and clandestineness are the words, until such time that we work out the setbacks to filter it to a sane conclusion," said Rishi.

"And who's that boy?" asked Rishi to no one in particular. "Ah yes! I recall. Anil! His father will be the first to put us on the dock for an uncalled cross-examination. His incessant probing drives me crazy and the absence of appropriate answers will make it hard for us to face his inexorable volley of questions. Do you remember the fuss he generated when we requested his son's presence in the Institute for an interim period? With a pretence of solicitousness and elusive references to his son's recovery we did succeed in cutting through his reluctance, but his terms spoilt our sweet moment of triumph. They were so ridiculously atrocious that for a moment I was tempted to walk out from there. But official etiquette demanded otherwise."

"I recollect," Diane affirmed, "but don't forget the gentleman's state of mind. He had just lost his wife and his son had a defect in his eyes. It was a stressful time for him. But he did agree, didn't he? You have to appreciate him for that. Anyway, don't worry; I'll handle the incognito aspect. The word 'Trust' has been ingrained into every employee of the Institute. The scale of confidentiality on this project is so high that besides the 4 of us, nobody has even been apprised on the UFO that we sighted a few days back, at least not yet. Except Steve and Ayaz. But they can be trusted. But that's not what worries me. It's the whereabouts of these boys that's unsettling me. What about you Alex? Any suggestions? What makes you so silent?"

Alex was doodling on his scratch pad absorbed in the pondering of the twists and possibilities of the most recent event. The shock had paralysed him, temporarily impairing the coherence of his thoughts. "I can't get over this. How could the boys just vanish into thin air?"

In the meantime, Sridhar, having arrived at his own conclusions, stood goggling at the tranquil airspace which only a few hours ago had threatened to plunge the world into an upheaval. He inhaled and exhaled deeply to conciliate his muddled mind, and temporarily retreated into a cloister of thoughts. He now knew for sure that his judgement of the involvement of an alien in the disappearance of these boys was unerring. His experience only reiterated it.

The sighting of the 8-gon shaped UFO had made him extra vigilant, but what happened the night before was unexpected. Not just unexpected but unbelievable! He reconned the sky looking for hostilities but found none.

It was difficult to digest that the entire sky had turned so startlingly opaque, as to make it impossible for him to see even a distance of 3 inches away from himself. And he was certain that the visibility had eroded severely only in this part of the sky. The news would have made headlines all over the globe, otherwise. Only one local newspaper had remarked on how pollution was causing the clear skies to appear grey even during the best of weathers; but that mention too was trivially relegated to an insignificant corner of the third page. Obviously, no one appeared to have sensed anything inappropriate.

He noted with satisfaction, however, that Rishi had treated his comment with a mark of respect. And now Diane had cautioned everybody against speaking to 'The Press.' That was a good thing. Too much of publicity at this time would be detrimental to his own personal research. He had vowed to keep a watch. He suspected that something similar would happen again, probably a few days later and the boys would be back to where they belonged. He kept his thoughts to himself, though.

He remained acutely receptive to the slightest of changes. His buoyancy, however, was undermined by the thought that when something extraordinary did happen, the incident would be insulated by the denseness of the atmosphere, making it impossible for him to observe the sky with a clarity he desired. He could only keep his fingers crossed. And the images? Well! Would there be any? He wondered tartly. But this time around there had been a notable difference in those pictures, if one could call it so. Instead of a slight hazy shade of red spread across, there had been a blue one. Well! Not exactly blue, but a mixture of blue, indigo and violet.

***

Mr. Varma looked at the message. It was amusing how the exigencies of life influenced one's decisions. Not long ago he had thought that the key to a peaceful existence lay in one's own hands. But the variables that had cropped up had forced him to think otherwise. Life was no more a straight road to a destination of one's choice. It was more like a jigsaw puzzle that one had to solve by picking up the right pieces to complete a picture.

It had been 2 months since both Diane and Alex had called on him, and now they had requested his permission to visit him again. He was glad. It was pertinent to have another round of discussions with them.

When the boys had first reappeared, they had been abnormally calm. There had been no sign of their aggression that had always been a cause for complaints in the immediate area, in the past. The neighbours had been astounded by their behavioural change and had even commended them for their impeccable conduct. The ethereal calming effect they seemed to be having on everybody they encountered was confounding, to say the least. Initially, Mr. Varma had been discombobulated, but with so many incidents documenting their transformation, it became increasingly clear that their healing touch was for real.

Their reappearance had been inoculated with something unearthly and halcyon. Insouciant in surmising that they had been transported to the heavens and blessed by angels, post their disappearance, he was lulled into thinking that things would remain this way until eternity. But all good things didn't last forever. The 'touch' had begun blunting. Contemporaneously so was the phosphorescence. Its intensity was now a pale shadow of its former self.

And now it unhinged him that their makeover was a total freak... an isolated case, a chimera. He could see and feel the change, a change that could undeniably be described as unpleasant.

In comparison to their pre disappearance days, the boys still appeared calm, but this façade of calmness was sometimes, if very rarely, punctuated with mild cases of aggression. The erratic outbursts seemed to be coinciding with the dimming of the diaphanous and waiflike luminescence that had become an inseparable part of these kids. The vicissitudes were piecemeal, constant, and subtle, suggestive of the pre-disappearance days of the boys.

He was secretly relieved when it was conveyed to him that Diane and Alex intended to pay him a visit soon. They were the only ones who knew this hush hush secret other than his wife and him.

The screech of the brakes informed him of their arrival. As both Diane and Alex, once again, made themselves comfortable in the artistically decorated living room of the Varmas, Diane noticed an almost imperceptible shift in Mr. Varma's otherwise readable mood, although his tone belied it. Apparently, something was disturbing him.

"So, how have you been doing?" Mr. Varma asked offering a strong grip in the form of a warm handshake.

"Fine, fine, thank you," replied Diane magnanimously, while Alex smiled in agreement.

An awkward quietness followed.

"I imagine you have a number of questions to ask me," Mr. Varma initiated the conversation with a sweeping gesture to put them at ease.

Diane nodded, "Where are the boys? And, how are they?"

"They're fine." replied Mr. Varma. "Slowly reverting to normal."

"Reverting to normal?" asked Diane finding it hard to hide her astonishment.

Mr. Varma dawdled. There was a lot he wanted to say, but couldn't. He chose to restrict his speech, "Exactly what you heard. The behaviour of the boys, at your first meeting, was a clever smokescreen for the original. It was antipodal to what they actually were before they were lost and found."

"I fail to understand," said Alex.

"If I remember right, I did mention to you, during our last conversation, that these two devils of mine were notorious in our neighbourhood. Not a day passed without a remonstrance; windows broken, flower beds trampled upon, smaller boys bullied and what have you. There were times when we were even forced to visit the police station for cases of violence. Then they got lost. Nobody was vocal, but I'm sure everybody in the neighbourhood must have been secretly happy," said Mr. Varma in a bid to explain what he meant.

He sighed, "When they returned, their metamorphosis had to be seen to be believed. It was almost as if God had held their hand and taken them to Heaven to transubstantiate them into seraphs. Some even claimed that the angelic look they seemed to be sporting, since their return, had a healing touch. I was grateful to the Almighty. But only yesterday I received a complaint - a whinge that was reminiscent of old times. I believe they kept chiacking this old man who lives down the street. His walking stick was hidden to stop him from his evening walks."

"What makes you think they were responsible?" asked Diane.

"We found the stick in the boys' bedroom," confirmed Mr. Varma.

"It's just an innocent leg-pull; so expected of young growing boys." said Alex. "You should take it with a grain of salt."

"You would not have thought so, if you had to be at the receiving end," justified Mr.Varma. "And most people would agree with me. It may have been an innocent prank, but pranks are pranks, and we, as parents must be held culpable for children's behaviour. I've begun getting glimpses of their old selves. The didoes are isolated but they are becoming more and more frequent."

"Where are they now?" asked Diane.

"At school. So far, there's nothing adverse from there. I hope it remains that way."

"We have come here to see them," said Alex. "Wanted to reconfirm our last experience with them."

"Your last experience with them?" Mr. Varma sounded perturbed. "But you hardly spent any time with them."

"Agreed," Alex replied, "but even those few moments left us a lot calmer. Diane even got rid of the ache that she had developed consequent to her meeting with boys who had 'blinking' eyes."

"I'm glad you broached this. Similar statements are being made here. The older people keep requesting me to send the boys over for a while. They feel that the 'eyes' of the boys act like a balm for their aching bones.

It is almost as if the boys have been beatified with the ability to heal. My wife who suffered from chronic arthritis in the past has not complained too, ever since their return. But now I am not sure."

"Why?" asked Diane, in an indecorous tone.

"They seem to be losing it," said Mr.Varma. "I'm referring to the healing touch. Only last night, my wife complained of a slight pain in her knees, although it was not as severe as it used to be. I've been putting 2 and 2 together, and I have concluded that the key to this mystery appears to be the fluorescence of that weird blue light that is visible just above the right ears of the boys. There appears to be a correlation between the glow and a change in their behaviour."

"A correlation?" asked Alex.

"I have been unfailingly observing the radiance; it has begun dimming. It was bright the first night I saw it but now it appears to have paled into insignificance. The miraculous touch of the boys appears to be waning in proportion to the diminishing of its intensity. I fear, it may soon disappear and when that happens, I'm cocksure I'll start receiving complaints like before. How I wish I could transport them back to where they had disappeared! I assert with certitude that there are angels living there."

Diane laughed a small laugh, "Esoteric indeed! I wish we could unravel it; but we do not know from where to start. These stray fragments need to be fitted comprehensively but they all seem so disoriented that I am beginning to wonder if they belong to a single picture at all. There was a time when we spoke of missing boys returning with evil eyes that cause the death of females, and today we are discussing 'eyes' that heal. What a contradiction!"

She looked at Alex seeking his asseveration. He nodded his head in assent.

When the boys returned from school that evening, they greeted Diane and Alex as respectfully as they did at their first introductory meeting. They looked angelic enough, but Diane thought she intuited a tinge of sarcasm in their politeness. She was not sure, though. Perhaps, Mr. Varma's statement had influenced her perception.

They spent that night at Mr. Varma's residence. They agreed that the glow had faded. It flickered on and off like a flame that would soon die away. But... Diane's sense of the unspecified heightened. She sniffed as inconspicuously as possible. There was a decided familiar tang. She couldn't quite place it, but it reminded her of something bittersweet. She searched her memory, but it failed her.

The next morning as they walked towards the nearest airport for their return flight to the Institute, Alex stated, "This colour blue? I'm beguiled by it."

"So am I," replied Diane. She hoped a miracle would happen and all her doubts would be dispelled with a shake of a magic wand. But what she was experiencing was real, and magic had no place in it. God! When would they solve it all?

* * *

At the Institute, Rishi jokingly advised Diane, "Have another discussion with Steve. You never know; he may come up with some explanation for the colour 'blue' too."

All of them broke into laughter. Alex injected the moment with seriousness, "What is enigmatic is not just the colour but its effect. Imagine Mr. Varma telling us that he wished he could send the boys back to the place from where they returned, only so that they would be 'good.'"

"Yes," said Diane. "I was dumbfounded when he said that. I gave no reply but somewhere at the back of my mind was a wishful thinking. If there really did exist a place where such goodness was found, then it would not be a bad idea to send all human beings out there so that they could transform themselves into angels. 'Earth' would then be 'Heaven.'"

"Wishful thinking! So like a woman to think so," said Rishi.

"What she says does make sense to me," said Alex emphatically. "There was an understated change in the boys' behaviour, this time. They appeared polite but insolent, unlike the last time when genuine innocence oozed from their attitude. Besides, we personally observed the diminishing light. I'm frustrated that I can't place my finger on it."

Diane put an end to the conversation by directing a question to Rishi, "By the way, have the boys been found? Have they returned?"

"No," replied Rishi. "Sridhar tells me that if his observations are anything to go by, then they are likely to return when we least expect them to. Let's see how his reconnaissance pans out."

* * *

After consecutive uneventful nights, Rishi, Alex and Diane decided to go home; but not Sridhar. As a part of his now-normal routine, he opened the window and looked out. The magnificence of the night as always left him awestruck. The stars overhead had arrayed themselves in a distinctive constellation.

Had he been a poet, he would have written a soulful poetry on the picturesque scene, but he was not; had he been an artist, he would have sketched a masterpiece, but he was not, although as a child many thought he was good at it. So, all he did was to appreciate nature in all her glory and hope that she would spring some surprises.

It had been quite a few days since the boys had disappeared, and from what he had seen and observed in the past, the boys should have returned by now. But they hadn't. The finale was taking a longer time. Of course! There was an instance of the twins who had reappeared after a longer gap than the usual. But this wait was even longer. Was it because 3 had gone missing? Perhaps the duration of disappearance and reappearance was in proportion to the number of people involved. Sridhar had conveyed as much to Rishi when he last spoke to him.

"Let's wait a while longer," Sridhar had said, "if nothing materialises, we can think of a new course of action."

Though Sridhar himself had not been too convinced, he was relieved that Rishi had accepted his suggestion. But with nothing untoward happening for several successive nights, he was on tenterhooks.

Crestfallen as he was, he persisted with optimism and continued to keep a close watch at the atmosphere round him, making sure that the pantry boys provided him with fresh coffee every 3 hours, to prevent him from falling asleep. He had to be alert. The unexpected always occurred when one least expected it too. He would start distrusting himself only after the 'expected' failed to happen within the next fortnight.

The days at the Institute had changed ever since the boys were reported missing. If 3 boys could disappear into thin air without a trace, there probably was some unearthly creature stalking the premises. The guards were extra vigilant and more of them had been deployed even at such places which were not considered strategic.

Rishi spent most of his time with Sridhar. He, along with Alex and Diane lingered at Sridhar's workstation, and scanned through the daily images captured by the CCD searching for incompatibilities. Although they appeared normal to others in their behaviour, there was no escaping the fact that the mysterious disappearance of the boys had unnerved them. They had nothing to offer to the world if questioned on their disappearance.

Uneasiness pervaded the entire Institute. The only one who appeared calm was Sridhar. He knew things would happen only when the sky started changing into unearthly colours, and that had not come to pass so far. When it did, the boys would return. Or would they?

His eyes traversed the clear sky for hints. By his own estimation the boys should have returned by now, but they hadn't. Way too many days had passed. The tension drew upon his nerves. Although he had confidently prophesied the miraculous return of the boys who had slipped behind a gossamer of nothingness, he now found himself wavering in his own credence.

He gave another look outside and was about to pull down the blinds when without a warning, the sky turned black and grey with a hint of deep blue, pronto. The atmosphere around him became impervious and dense; it was evocative of the night when darkness had neutralised all light, to steal the boys away. Visibility eroded as on that fateful night; he could scarcely see a thing even 6 centimetres away from him. His compressed excitement intensified his senses but failed to ignite his mobility. He stood rooted to the spot. Then the sky cleared, almost as if nothing had happened. Without losing a moment, he dialled Rishi's number. "I think you should come to the Institute, immediately," he said.

"Has something unnatural eventuated again? Nothing scandalous, I hope?"

"The same thing that happened on that cabalistic night."

"You mean the opacity? I'll be right there," said Rishi hastily getting into his semi-formals. Navigating his car at lightning speed, he reached the Institute, but by then, the sky had cleared and the opaqueness gone. "Oh my!" sighed Rishi, I've lost the moment."

"Yes!" said a visibly vivacious Sridhar. "But that's not why I called you. Do you recall that night? The boys had disappeared then, and tonight being a replay, either the boys have reappeared or someone else may have been lifted from the premises. Quick! Let's check!"

"God! You are right!" exclaimed Rishi. "Why did I not think of that?"

They rushed and sure enough, all the 3 boys were found sleeping soundly on their beds. "Astonishing! Should we wake them up?" asked Rishi, unable to contain his excitement.

"Not now," cautioned Sridhar. "They appear more sedated than asleep."

"Perhaps, you are right," conceded Rishi. "Let's just switch off the lights. The darkness has a soothing influence over a person's sleep. Besides, the lights too need to rest. They have never been switched off since the boys went missing, and you can give me a second by second account of what occurred, while we walk towards my cabin."

Sridhar switched off the room lights as instructed and then immediately let out an involuntary gasp, "Look!"

Rishi turned. His exclamation was even louder. For one long moment both stood immobilized. A strange azure gauzy fluorescence was visible above the right ear of the boys. It had been invisible when the lights were on.

"Diane must see this. She kept mentioning a cerulean luminosity that she had seen in Mr. Varma's house. Ayaz had even captured it on camera. But there is a world of difference between the images and the actual glow," said Rishi in a whisper.

***

The next night when Rishi, Alex and Diane tiptoed into the boys' enclosure, Diane was unable to contain her enthusiasm. "Yes!" she said. "This is exactly what I was referring to. Isn't it Alex?"

Alex shook his head disbelievingly, "The rapidity with which everything's happening will turn me insane."

"Telling me!" said Rishi. "I am flummoxed. How do we react?"

"By pulling out tufts of hair from our head," replied Alex.

"Let's call Ayaz," said Diane. "He'll have much to say and much to do."

"I'm relieved," said Rishi 2 days later. "Now that the boys are back we should arrange to send them back. I've run out of excuses. Anil's father has been telephoning constantly and so has the Master of the boy, Gattu. I believe he's a conscientious worker. He's finding it hard to find a suitable replacement. Ayaz is the only odd one out. He's agreeable to keeping his protégé with us."

"No! Not now!" protested Diane. "We have to keep them here for another 3 months, at the least. There's a parallel between Mr. Varma's version and ours. During our last visit, he kept recounting the correspondence between the waning of the bluish glow and the behaviour of the boys."

"Yes," intervened Alex. "The boys should be kept under observation for a while. Both Diane and I noticed the change in the incandescence. Just inform the guardians that the boys will be healed when they return. That'll pacify them. Have you noticed how the boys have stopped blinking unnaturally? Besides they appear ethereally serene and calm. It appears to me, that the boys have all visited a place of tranquillity during their absence from the Earth."

"But where?" asked Rishi in exasperation.

"That's something we have to find out," replied Alex.

"Mars?" thundered Rishi in exasperation. "But that's supposed to be hot, isn't it?"

"No idea," said Alex. "All I know is that I'm fatigued. The clamour has been too much to digest. I hope the morrow does not spring more bombshells. I am sure to suffer from a cardiac arrest if it does."

In the meanwhile, Sridhar was disappointed. He had kept vigil all nights in anticipation of a sensational discovery and all of it to no benefit. His prediction of a change in the sky had been correct, but the abruptness of it took him by surprise. He had been powerless in such a scenario. No magnitude of preparedness would have helped him discover what really transpired under the screen of such fogginess.

It was obvious that the opacity was a camouflage for the transportation of the boys. Unmistakeably, someone or something was responsible for it. But who? Dammit! His inquisitiveness was further aroused, but he knew not how to get to the bottom of it. There were no warrantable clues to guide him on the path to discovery, but he would persevere. Someday his resoluteness would pay off.

The following night, he didn't stand by his window. Instead, like before, he walked out into the open with a grim determination. He looked at the vast expanse above, searching for some remnant of the previous night's atmospheric disturbance. There was none. The sky was clear except for an occasional fleck of cloud that appeared to be flirting with the moon. The stars sparkled as usual. Did last night really happen or was it a dream? He shook his head disbelievingly. The presence of the boys was evidence that it had.

The time had come to immerse himself in those huge hardbound books that were collecting dust in his library. Perhaps he would find a hint there. As for now, he had to be satisfied with looking up at the sky and trying to figure out if he could spot anything outlandish. A new assignment had now been added to his existing work profile - to search for alien life.

***

CHAPTER XIX

"We are ready," claimed an enthused Athenia.

Phoebia looked at her, "Let's review. The calculations? Have you perfected them? The target? Have you focussed on just the Martian workshop, and not the whole of it? The missile should in no way be diverted from its intended target. Have you considered, reconsidered the consequences of the detonation? Unwarranted contretemps would be corrosive. We want no sitch that will tilt the consummate egalitarianism of Space."

Athenia's reaction revealed her priorities, "Rest from worry! Precautions have been taken. The moment the fine-tipped missile touches the 'Martian Space Experimental Laboratory,' the pellets will explode unilaterally destroying the 'testing workstation' with all its contents, leaving behind nothing but a black space. The quickness of the attack will leave the Martians with no time to react."

"And the aftermath?" prodded Phoebia.

"Since the missile travels several times faster than the speed of light, disintegrated particles are bound to be thrown into outer Space once it hits the Martian soil. Some fragments may enter the Earth's atmosphere at a speed ranging from 10 to 70 km/

sec that would soon decelerate to a few hundred km/ hour by atmospheric friction. Hopefully these will be small, just a few milligrams each, by weight. If bigger shards in the range of 20–40 metres in diameter reach a densely populated area on the Earth's surface, then the possibility of casualties cannot be ruled out. But that's a risk we have to take if Earth has to be saved in the long run," explained Athenia.

"Has anything been done to minimise the risk?" countered Phoebia.

"Yes," replied Athenia. "The explosion has been configured to largely eliminate such an eventuality. The flotsam would either splosh into the water or touch barren land. We'll have reason to rejoice if it squelches into the ocean. If it doesn't then there would be some damage."

"Damage! That's exactly what I wish to avoid," emphasized Phoebia.

"Worry not!" consoled Athenia. "It would be limited to the caving of the land where the sliver meets ground. The result - craters due to the sheer impact of the speed of the splinters."

Phoebia sighed in agreement. "Your words encourage me. I guess we have to be realistic. I take solace in the thought that our action is purely placatory. The conduct of the Martians is a matter of grave concern and sufficiently serious to warrant action from our end. Everything else pales into insignificance."

Athenia concurred with Phoebia. The Space area had been reconnoitred earlier, and she was satisfied with the grounding. The combat was one-sided, and warranted an immaculate perpetration. She was confident.

As she steered the Venusian Spacecraft to the vantage point, nostalgia beleaguered her momentarily. She recalled the last occasion when she had been involved in a similar attack. Things were markedly different then. It had not the slightest hint of despoilment, not even a peripheral one. All that was expected, then, was the darting of a coded message that warned of dire consequence if Earth was destroyed. The communiqué had worked. There was peace in the Cosmic Universe for a long time, thereafter.

Why? She was not sure. Did the Martians really fear them?? Or was it just an impermanent placation to throw them off their track? Well! The reasons didn't matter. What mattered was that the frosty advisory had been respected. But now she deduced aptly that it had just been a calm before the storm.

She looked around. To avoid glitches at the nth hour, an immaculate search of the area of attack had already been undertaken during the recce. Unpleasant shockers were not her cup of tea. She inspected the tiny disc approximately the size of a drop of rain that fell on Earth; it sat innocently ensconced in the front end of the Venus Max Spear. It had the power to destroy the entire Martian Planet but had been coded to have just a partial impact on it. "Well! Nobody likes destruction, but if this is the only language that the Martians understand, then we have to speak it."

***

"Utterly unpropitious! Things couldn't have got worse for us."

"The referenced subject? If I may ask," asked Maxus.

"The chips. They have ceased to function. There are no signals... not the slightest. The way I see it, it would be wise to abandon the entire project. There have been way too many roadblocks, and outmanoeuvring them has been irritatingly vexing. Sabotage! Someone's trying to sabotage our plans," vociferated Pontus. His voice had an edge of ignominy. It duplicated Maxus's unsettling thoughts precisely.

It was incensing to see a project crumble because of the machinations of the unseen. They had been outwitted - literally outfoxed, once again.

"That's a bad omen! Who do we hold responsible? The Venusians? Or, a traitor in our midst? Anyway, lamentation is futile after the damage has been done, but it should act as a guide to obviate similar ambuscades in the hereafter. Focus should be shifted onto the new modified Project which is all geared for take off," said Maxus in an effort to soothe the ruffled feathers of Pontus.

A reverberatory thunderous explosion echoed across the dialogue.

Caught off balance, Maxus and Pontus looked at each other.

"Did you hear what I just heard?" asked Maxus.

"Yes!" Pontus replied. "And it doesn't sound like good news to us. Let's check."

They both flashed towards their workstation - the most erudite piece of work, absolutely state-of-the-art with no parallel in the entire Cosmos. There was nothing left of it. Not anymore. All they saw was a deep black hole where once stood their most sophisticated work of 'Science.'

The exploratory fine-tipped Venusian missile had done its job. And how!

* * *

The News room was abuzz with excitement. The Research Institute had only just released a Bulletin warning the public of the possibility of a huge meteorite hitting the surface of the Earth in the next 24 hours. It hoped to influence its course of travel to avert disaster. For the first time in the history of mankind, sophisticated technology had made it possible to monitor its travel in Space and relay a minute by minute frame, to thousands across the globe.

People were glued on to their television sets to watch the momentous event. When, finally, it did enter the Earth's atmosphere, it fell into the ocean. The squish was a huge one. Had it touched the hard surface it would have braked joltingly with a clatter of sparks and rendered everything surrounding it beyond identification!

"It's a miracle," concluded the News reader. *"Thanks to the timely warning of the Institute and their constant efforts to divert its entry to the sea through their revolutionary technology, there are no causalities. We are proud of our Scientists."*

Rishi looked at Sridhar, "Did we really deflect its routeing?"

Sridhar allowed himself a slight smile. His constant vigilance had helped him reach a mind-boggling conclusion which was best kept to himself; at least for today. He'd be considered non compos mentis if he did. He knew with certitude that his recent bizarre observations had been correctly assessed by him, but this was no time to discuss them. He had yet to support them with deep-dyed corroboration. He would reveal his findings at the appropriate time. He knew his moment would come, and it would be exclusively his. Until then, he'd wait and watch and work.

He had been unfailingly precise in his scrutiny. He had come to notice that the discolouration of the sky by an outlandish red shade spelt disaster while an unrecognisable bluish hue meant that an anticipated disaster would be averted by a friendly alien. In the excitement of watching the trail of the meteor live on the screen, very few had noticed a concurrent, discernible change in the sky. The visibility had eroded severely and then cleared as swiftly leaving in its trail, albeit for a few blinking seconds, a mixture of colours of blue, indigo and violet. He knew then, that they would be protected.

"'Mars' and 'Venus,'" Sridhar muttered to himself.

"Did you say something?" asked Rishi.

"Nothing, nothing really!" whispered Sridhar enigmatically.

"Are you sure?" asserted Rishi.

Sridhar did appear strange. Unusually strange. Was he hiding something?

* * *

CHAPTER XX

SIX MONTHS LATER

Tying up the loose ends wasn't easy, but they had done a 'supercalifragilisticexpialidocious' job. War it was for them! A war that was brilliant in conception and mettlesome in triumph. The finis was simply coruscating.

"Things seldom turn out the way we intend, but it has. The Cosmos has attained *nirvana*, but for how long is the moot question. Times like these are priceless no matter how high a cost one must pay for it. It's a pity the Earthlings are oblivious to the aura of persistent disaster hovering abstractedly over them. Do you think they'll ever know that we protect them doggedly from the sinister manipulations of the Martians?" queried Phoebia.

Athenia laughed. "Never. They are busy with their Martian ambitions. Venus is not in their orbit of interest. But diversions such as these alleviate the otherwise serious drudgery of our monotonous existence. Carte blanche! That's what we possess when it comes to testing our expertise and reinforcing our knowledge - the last exercise, for instance...That was the only time we made use of our 'GOSSAMMER MASK.' In fact, we made use of it twice; once to bring them here and the next to transport them back to Earth again. It was necessary as there were 3 of them to be hauled at once, but I am glad that it was an experiment that succeeded, and perhaps just as well. Otherwise my resilience would have been undermined in the knowledge that our long laborious study was of no avail."

"Ah yes!" Phoebia acquiesced. "The allusion is to the airlifting of those 3 male human species, isn't it? It was made easy by our advanced opaque gas. It was a first, and a great success. Hopefully, there'll never be another occasion to use it in the future. The consequences are disastrous."

"Too disastrous," asserted Athenia. "Like all our prime inventions, this too comes with a host of demerits. Constant use of it can weaken the gravitational pull of any planet resulting in weak gravity or no gravity at all... and what would happen then? Well! Well! What an amusing sight it would be to see all living creatures floating in Space. That would be funny indeed!" said Athenia.

"Funny? Now that's what I call an UNDERSTATEMENT," countered Phoebia. The laughter that echoed was loud and long enough to cause tremors on Earth.

* * *

"Was that a quake?" said Rishi as the cup in which he held his coffee, trembled.

"No," replied Alex as he sipped his hot coffee slowly and deliberately, staring at his fingers which too quivered simultaneously. "It's a reminder."

"Reminder?" asked Diane.

"Yes," stated Alex. "A reminder to not take things for granted. It's been 6 months now since we last heard about 'blinking boys,' isn't it? I feel strangely uneasy. It's been quiet for far too long. Do you think it is the lull before the storm?"

"No pessimism, please," cautioned Rishi. "Never mind the fact that we were never the paramount characters of the final tableau, yet it's amusing how everybody attributes all the credit to us. We have only just started savouring the fruits of our labour, and revelling in the eulogizing. Let's enjoy it while it lasts."

"But boy! Am I glad that we kept the boys with us for 3 months," stated Diane. "The bluish glow was phenomenal, but what was even more flabbergasting was the way it dissipated and totally disappeared at the end of the period. I guess we will never get to know anything about it."

"And what was it that you said during the interview?" asked Alex looking at Rishi amusingly.

"Don't ask me. I was dumbfounded when they said, "So, Mr. Rishi, how does it feel to be the 'Messiah' of this Modern Era?" A Messiah? The question knocked the stuff out of me, but I managed to regain my composure to give a poker-faced reply."

"I liked your answer," added Alex. "It's the team and not I, and if we had to do it all over again we would willingly do it. That's the spirit, man! Though I'm still trying to figure out what is it that we'll have to do again?"

Laughter pervaded the air.

"In retrospection," said Rishi, through the laughter, "I realise that I may have sounded bombastic, but at that time I was not sure if they were referring to the diverted flight of the meteor or the sudden cure of the 'blinking' boys," said Rishi.

"Both I guess," said Alex. "It amazes me how the simple passing of time can alter situations. There has been no news of that crazy blinking ... or has there?"

"No," replied Diane confidently. "Time indeed can render so many problems meaningless. Anyway, all's well that's end well. We can now go back to doing what we do best - concentrating on our actual 'Research' instead of being diverted by all else. I expect we should be content, now that we have seen and heard the last of those wearisome inexplicable appearances and disappearances. For a change, I smell the spring in the air."

Rishi agreed enthusiastically, "but..." his voice tapered into an expression of puzzlement.

"But?" Diane quizzed him curiously.

"Someone must have solved the enigma. Who? Although we may not know what we need to know, I'm sure someone does. Who, I wonder?" mulled Rishi.

"Yes, who?" questioned Alex. "The accused is still on the run. We can rest, but only for a while. We aren't safe until the killer has been found and hung on to a noose. Is this the end? Or is this the beginning? We'll have to wait and watch. Like I said before, I am still not cured of the uneasiness. It continues to shadow the smugness of success attributed to the Institute."

* * *

Maxus looked uneasily at Pontus, "When is he expected?"

"Any moment now," replied Pontus.

"Are you prepared?" asked Maxus.

"Yes."

"I admire your confidence."

"I know their weakness," quantified Pontus.

"Their weakness?" Maxus appeared puzzled.

"Yes."

"And, what is it?"

"Several," specified Pontus, "but predominant is their 'EGO'; pamper it and all will be forgiven."

"But this is the second chance that has been given by them?" reminded Maxus.

"Don't worry. There'll be a third," said Pontus exuding confidence.

A sudden whirr lighted the Giant Celestial Monitor and interrupted their schmooze.

"PONTUS!" bellowed the loud low-pitched voice, "This is XENTILIOS! You have failed us... AGAIN!"

Pontus genuflected before the vague caricature that showed itself on the 'Cosmic' screen. The features were hardly discernible. He had never been able to identify the persona behind the booming voice that issued orders as if it commanded the whole of Universe. His own obsequiousness was fake - merely to assuage the anger of the voice behind the diktats. Did he fear him? No! He never would. After all, the ball was in his court. But he needed him. He was an easy conduit for information on Earth.

He had been measuring humans for a great length of time and knew enough of their nature to deduce that a petty gesture such as a mere bending of the head and knee would tickle their ego and incapacitate their reasoning. They were suckers for flattery. These men who professed no appreciable accomplishments rode roughshod over the rest of their clan merely on the strength of their draconian façade which cleverly hid their cowardice. The chip was a clever tool; an ingenious chicanery to wipe out those they feared.

"We almost succeeded," Pontus stated in a tone that was respectful and slavish. "But the cursed Venusians intervened at the inopportune moment."

"Stop fooling me, Pontus!" thundered Xentilios. "Ours was the script; yours - the technology; a technology that could have put even the fictitious Venusians at bay."

"We were caught unawares," countered Pontus. "Besides, they are not fictitious. They exist just as you and I do."

"I refuse to believe you," reiterated the voice. "Anyway, the problem's yours, not ours. We want results. It was clear from the very inception that while you handle the implant of the chips, we would use our own discrete machinery to monitor it at our end. To work in tandem, in a manner that would garner neither attention nor identification, was the intent. Where did it go wrong? And how? Was the chip modified? Was it tampered with? Was it rigged? I need no excuses. Only an answer. Now!"

"None of what you state is true," defended Pontus.

"Then why was it flawed?" demanded the voice.

"It wasn't," asserted Pontus.

"It was," roared the voice. "The entire store of genetic information was imparted to you, to enable you to disseminate resilient genes and build an unchallenged superior human race - second to none in all the 9 dimensions of intellect, youth, wisdom, strength, beauty, power, longevity, resilience, immunity and indestructibility. A human race that would sail past eternity and continue to rule Cosmos, incontestably."

The deep-toned voice paused. Pontus stifled a sneer, and continued to genuflect before it. The menacing uncommunicativeness was a temporary suspension.

The voice then spoke again.

"The chip was coded to decimate fragile females and render the weak males impotent. The job was assigned to you because of your superior technical prowess. You tricked us. Strong females were your targets and the males... Well! The less said about them, the better. It is clear now, that you no longer have the slightest understanding of what is expected of you. You have disappointed us once again."

"I'm sorry," apologised Pontus in a tone that was deceptively respectful. "The instructions were misunderstood. I'll recheck. The chip may have malfunctioned. One more chance is all I ask of you. The third and the last. You will get your heart's desire."

"So be it. One last chance, Pontus. An Earth year from now, we meet again. A failure will no longer be condoned. If that happens, you will be doomed. Remember what you have is just the copy. The original is still with us. Unlike the past, we will be tracking every movement of yours henceforth - any tinkering, and we will know. Once bitten, twice shy. Start working!" commanded Xentilios.

The voice was succinct, clear... separated by indistinct sounds that served as weighty pauses.

The giant Cosmic Screen hissed... susurrated... and whirred once again. The boom dissipated in Space.

"Phew!" Pontus heaved a sigh of relief.

Maxus then spoke, "So, this is the source from where you get your never-ending supply of knowledge on humans. I thought we had made a brilliant headway in our Research on the strength of our talent. How did you meet him?"

"Our paths crossed on a radiowave. He spoke of his desire, and I looked upon it as an opportunity. The story is long and is best kept aside for another time," declared Pontus.

"I'm willing to wait," accepted Maxus." But what is 'Xentilios' seeking from us?"

"You heard what he just said, didn't you?" countered Pontus. "He wants to cleanse the human race of its weak genes and create an exclusive one."

"Is that true?" asked Maxus

"Do you think it is?" countered Pontus. "All he wants are serfs. He wants us to shape the chip to build a world of followers who would do everything at his bidding."

"Is he alone in his pursuit?' queried Maxus.

"No," stated Pontus. "He has an entire conglomerate with him. This clandestine agglomerate is a feudal nightmare. They want to live, but not let the others survive unless they are willing to be slaves. They foolishly consider themselves to be superior and behave in a fashion so contrived as to make the others believe in their masquerade. They forget they are as mortal as the rest."

"Are they truly prodigious?" asked a visibly nervous Maxus. He had yet to get over the formidable conversation.

Pontus seemed to be unperturbed. "No, they are not," he said emphatically.

"But he appeared so," insisted Maxus.

"That was a façade," explained Pontus. 'Ego' and 'Unchallenged Power' pair up to make an intoxicating aphrodisiac. Out there, power is like a transposed tornado where the winds are arguably the greatest at the narrowest point. Critical decisions taken at that point by the very few who choose to call themselves the influencers have a cascading effect on the general proletariat. Their 'we do good' comportment is a mask for their flagitious designs - to destroy their own. They are eviller than we are."

"Why don't they employ their nuclear capability?' asked Maxus. "They have been working on it for a long time."

"That would maim them all, including the likes of him."

"Genocide?" prompted Maxus.

"That would build a psychological momentum among the proletariat who would quantitatively dislodge them from power. Besides, how would that help in segregating the wheat from the chaff? It is important to strain the genes to suit his desideratum. The chip has been coded to do that," stated Pontus with exceptional clarity.

"Then what exactly happened to the chip?" enquired Maxus.

"It was vandalized," declared Pontus.

"Vandalized? By whom?"

"By me."

"Why?"

"It was the best thing to do. What makes you think that I'll let a bunch of egoistic human fools rule the Cosmos? They have no fundamentals. They possess little or no ingenuity. All they do is to hoist their ambitions on fragile strings of luck that can easily be blown away by a gust of wind, any moment."

"But you just promised him."

"Promises are meant to be broken."

"Think over!" cautioned Maxus. "Your attitude should not develop into a troublesome competitive burden. 'He' has threatened us with dire consequences if we fail to bide his commandment. 'He' will wipe us out. Don't forget 'He' is the Master of the knowledge that he claims to have imparted to you. Most of what you know about humans comes from 'Him'. This has set in a spate of fresh doubts in my mind. Is their information reliable? What if it has been distorted to mislead us? Our dream of ruling the Cosmos, unknown to them would crumble down like a pack of cards in a matter of moments."

"The information may have been unbosomed by them, but it has been validated by *us*." Pontus gave a megalithic stress on the word 'us.' The script is theirs, but the technology is ours. Technology can do wonders. *It can destroy, it can build. It can impair, it can enhance. It can obliterate, it can create.*"

"So, what do we do next?"

"Hit them where it hurts the most."

"And where would that be?"

"The DNA!"

"The DNA?"

"Yes!" stressed Pontus. "The chip has already been pilfered by us. All that needs to be done is to deface it further. Our work station may have been destroyed - but not the prototype. It is intact."

From behind his ears, Pontus whisked out a replica, "Here it is. 'Mars' will always be the 'Master of Space.' Our attempts will never be obviated. Certainly not by those who have no unity among their own. What is it that the Earthlings say about resting on one's laurels? Well! Let them rest on theirs while we continue with our mission with an even greater frenzy."

Maxus looked at Pontus in awe. He had always admired him but now his admiration for him notched up further.

"When do we start?" Maxus asked

"This very instant," emphasized Pontus.

"Our sample?"

"He's been chosen."

"Who?"

"Our very first. He is young, smart, intelligent, resilient, resolute, strong, semi-schooled and above all vulnerable. He would be easily controlled by us. We'll crown him to destroy the world and annihilate the entire human species."

Pontus winked a wicked wink.

* * *

Grrrrrrrrrr... Gattu woke up to a sound he had heard, once before, while walking through the jungles. He sat upright on the narrow cot in the servant's quarters of the Master's house. He felt a familiar tingle as he wiped the heavy perspiration on his forehead with a broad chequered handkerchief.

Life had been a roller-coaster ride for him. His brief stint at the Institute had been enjoyable but there was no place like home. He had been cured of his weird blinking but not of the unfair accusations. Most people continued to look at him as if he was a devil in disguise. To them, he was the cause for a spate of female deaths including that of his mother. The Master, though, was very understanding. His generosity and kindness had touched the right chord in his heart, but it had not helped him rid himself of the trauma that had plagued him.

And now that sound! Was he imagining? The jungles had been left behind a long time ago. Then where did it originate from? He glanced at the shut window, walked towards it and opened it wide. It overlooked the bounteous acreage of the Master's estate. His bare feet that touched the cemented floor, felt colder than usual. It was night, and dawn was still a few hours away. As he peered into the darkness, memories and misgivings flooded through his brain, like huge raindrops hitting hard against a window pane. Far, at a distance, he saw a strange red light. He had seen it before but had never been able to identify its source. Curiosity got the better of him. He hurriedly slipped into his sandals, buttoned his shirt and walked out of the room. Letting himself out of the gates of the Master's residence, he found himself walking hypnotically towards it. As he reached closer and closer, it got brighter and brighter...

* * *